Working with Excel

Working with Excel

Refreshing Math Skills for Management

Priscilla Chaffe-Stengel and
Donald N. Stengel

First published in 2012 by
Business Expert Press, LLC
222 East 46th Street, New York, NY 10017
www.businessexpertpress.com

ISBN-13: 978-1-60649-280-2 (paperback)

ISBN-13: 978-1-60649-281-9 (e-book)

DOI 10.4128/9781606492819

Business Expert Press Quantitative Approaches to Decision Making
collection

Collection ISSN: 2163-9515 (print)
Collection ISSN: 2163-9582 (electronic)

Cover design by Jonathan Pennell
Interior design by Exeter Premedia Services Private Ltd.,
Chennai, India

First edition: 2012

10 9 8 7 6 5 4 3 2 1

Printed in the United States of America.

Abstract

Managers and analysts routinely collect and examine key performance measures to better understand their operations and make good decisions. Being able to render the complexity of operations data into a coherent account of significant events requires an understanding of how to work well in the electronic environment with raw data.

Although some statistical and financial techniques for analyzing data are sophisticated and require specialized expertise, there are methods that are understandable by and applicable to anyone with basic algebra skills and the support of a spreadsheet package. While specialized software packages may be used in a particular business setting, Microsoft Excel is routinely available on computer desktops. Managers who have been in the field any length of time may not be sufficiently familiar with the capabilities of Excel to make optimal use of its functionalities. Prior to undertaking a program to pursue executive training, managers who are refreshed with basic algebra skills and the capabilities of Excel will be prepared to develop a richer understanding from their more advanced work.

The primary foci of this text are (1) to refresh fundamental mathematical operations that broadly support statistical and financial equations and formulas, (2) to introduce work with equations and formulas in Excel spreadsheets, (3) to expand statistical and financial analysis with the programmed Excel functions readily available through the "Insert Function" toolbar button and the more advanced Data Analysis Toolkit, (4) to facilitate graphic representations of data, and finally (5) to create mathematical models in spreadsheets.

This text is a companion to a series of books that addresses the analysis of sample data, time series data, managerial economics, and forecasting techniques. Together these books will equip the manager and the student with a solid understanding of applied data analysis and prepare them to apply the methods themselves.

Keywords

Microsoft Excel, refreshing math skills, business math, statistical analysis, financial analysis, mathematical modeling

Contents

Preface

The Culture of Learning

The corporate world is a different environment than the academic world. Admitting one is not on top of, even controlling, important aspects of corporate business is an unacceptable practice in a board room. Therefore, to confess to areas that are not components of personal mastery can be deadly.

Managers who have been in the field any length of time may not be sufficiently familiar with the capabilities of Excel to make optimal use of its functionalities. Prior to undertaking a program to pursue executive training, managers who are refreshed with basic algebraic skills and the capabilities of Excel will be prepared to develop a richer understanding from their more advanced work.

The primary foci of this text are (1) to refresh fundamental mathematical operations that broadly support statistical and financial equations and formulas, (2) to introduce work with equations and formulas in Excel spreadsheets, (3) to expand statistical and financial analysis with the programmed Excel functions readily available through the "Insert Function" toolbar button and the more advanced Data Analysis Toolpak, (4) to facilitate graphic representations of data, and finally (5) to enhance understanding of the application of Excel spreadsheets in mathematical models.

As authors, we have considerable experience teaching courses in business statistics and applied mathematics for business and management. Often we find students struggle with these more advanced topics because of underlying weaknesses in basic mathematical operations and algebraic thinking. Attention to algorithmic detail is essential in constructing, using, and interpreting the results from mathematical modeling. Our intent in writing this book is to provide a supplement that develops mathematical expertise while learning and applying that expertise in real-world situations.

In picking up this book, you begin a different journey, a journey of learning, development, and expansion. Learning is a constructive activity. Learning can be exciting as new ideas forge connections to long-held understandings. Sometimes old ideas have to fall for new ideas to find fertile ground. Learning new things can take you into new territory, which can be a humbling experience. Learning is hard work. But the learning process can be made easier by recognizing the insecurity it can easily muster along this path. Monitoring the emotional character of learning can be helpful along the journey.

CHAPTER 1

Working with Numbers and Equations

Managers and analysts routinely collect and examine key performance measures to better understand their operations and make good decisions. Being able to render the complexity of operations data into a coherent account of significant events requires an understanding of how to work well in the electronic environment with raw data.

Although some statistical and financial techniques for analyzing data are sophisticated and require specialized expertise, there are methods that are understandable and applicable by anyone with basic algebraic skills and the support of a spreadsheet package. While specialized software packages may be used in a particular business setting, Microsoft Excel is routinely available on computer desktops. Prior to renewing familiarity with the capabilities of Excel, managers should be refreshed with basic mathematical and algebraic skills to be prepared to develop a richer understanding from more advanced work.

1.1 The Magnitude of Numbers: A Quick Review

Understanding magnitudes of numbers is an important starting point. Which is greater: 0.014 or 0.01? −1.96 or −2.01? On the surface, these questions may seem trivial, but weighty decisions made in statistical terms may well rest on the appropriate comparison.

1.1.1 Evaluating Numbers by Hand

To evaluate positive numbers by hand, we align the numbers by their decimal points and then complete any digits missing to the right of the decimal with a "0," as shown in Figure 1.1.

Recalling the place values of decimals shown in Figure 1.2 allows the straightforward comparison of the two numbers: 14 thousandths is larger

Figure 1.1. Aligning decimals at their points.

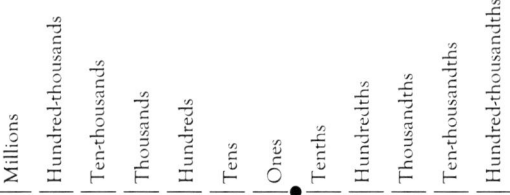

Figure 1.2. Establishing place values of decimals.

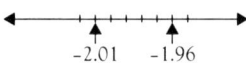

Figure 1.3. Number line comparison.

than 10 thousandths. In mathematical notation, we can conclude that $0.014 > 0.01$.

To evaluate negative numbers by hand, we may need to repeat the process shown above. In a final step, however, we need to reference a number line, remembering that smaller numbers appear to the left and larger numbers to the right, as shown in Figure 1.3. The further a value is away from zero, the more negative the value is, and the smaller that number is.

So, -2.01 falls to the left of -1.96 and is smaller than -1.96. In mathematical notation, we can conclude then that $-2.01 < -1.96$.

Notice in both cases that the inequality sign points to the smaller number and opens to the larger number. In fact, directional symbols and some of their meanings are shown in Table 1.1.

1.1.2   Using Excel to Evaluate Numbers

We can evaluate numbers quickly with the use of Excel. Simply type each value to be compared into a cell. Arrange the cells contiguously either in a single row or a single column. Highlight the cells containing the numbers, select "Data" and then select "Sort." If sorted in ascending order, Excel will place the smallest number first to the greatest number last. If sorted in descending order, Excel will place the greatest number first and the smallest number last.

Table 1.1. Inequality Signs

Symbols	Meaning
$<$	• Less than • Fewer than
$\leq$	• Less than or equal to • No more than
$>$	• Greater than • More than
$\geq$	• Greater than or equal to • No less than

1.2   Order of Operations

The order in which mathematical operations are performed is significant. Different answers can be had from the same numbers and the same operations conducted in different order. For example, $(3 + 1)^2$ is 16, but $3^2 + 1$ is 10, and $3 + 1^2$ is 4. We adopt common conventions to make sure we are all referring to the same procedures conducted in the same order so we can arrive at the same answers. When we deal with formulas and equations, the order of operations involved in evaluating the formula or solving the equation can be critical, whether the formulas and equations address significant financial applications or complex statistical analysis, as we will see later in this chapter.

The order in which mathematical operators are activated is:

1. **P**arentheses,
2. **E**xponents,
3. **M**ultiply and **D**ivide, whichever comes first left to right, and
4. **A**dd and **S**ubtract, whichever comes first left to right.

We have highlighted the first letters, **P-E-M-D-A-S**, to trigger the mnemonic device you probably learned years ago: **P**lease **E**xcuse **M**y **D**ear **A**unt **S**ally, capturing the order in which operations should be conducted.

1.2.1   Parentheses First

Parentheses establish priority. Information grouped within parentheses should be evaluated first. When parentheses are immediately preceded by a number, multiplication is implied. So $3 + 2 \cdot (5 + 7)$ means that you add 5 and 7 first, then multiply that sum by 2, and finally add 3 to get the answer 27. If the elements grouped by parentheses are algebraic variables, as is the case in the expression $5 - 2 \cdot (2x - 3y)$, where you cannot combine the terms inside the parentheses, then you must eliminate the parentheses by using the distributive property. If operations are conducted in Excel, however, a multiplication sign, indicated by the sign *, must be inserted between the number and the parentheses.

The Distributive Property of Multiplication
Over Addition or Subtraction

Algebraically, when the sum of two numbers is multiplied by a third,

$$a \cdot (b + c) = a \cdot b + a \cdot c$$

the multiplier a can be individually distributed to each of the numbers summed inside the parentheses, b and c.

So the expression $5 - 2 \cdot (2x - 3y)$ expands with the multiplication of -2 times the binomial to achieve $5 - 4x + 6y$. No further simplification is possible unless we know specific values for x and y to substitute into the expression and evaluate it. Sometimes the distributive property can be used in reverse, in a sense, to remove a common factor. For example,

$$597 \cdot 439{,}845 + 403 \cdot 439{,}845 = (597 + 403) \cdot 439{,}845$$
$$= 1{,}000 \cdot 439{,}845 = 439{,}845{,}000.$$

Take care with negative signs that appear in front of parentheses. If the multiplier is negative, as is the case in the expression $5 - 2 \cdot (2x - 3y)$, the negative number is distributed to both terms inside the parentheses. Table 1.2 contains a quick reminder of multiplying positive and negative numbers.

So $-2 \cdot 3 = -6$, but $-2 \cdot (-3) = +6$. When there is just a negative sign in front of parentheses, for example $- (4p + 5q)$, consider the expression as $-1 \cdot (4p + 5q)$ and use the distributive property to multiply -1 times both $4p$ and $5q$, to yield $-4p - 5q$. In general, if there is an even number of negative quantities multiplied together, their product is a positive

Table 1.2. *Multiplying Positive and Negative Numbers*

Symbols	Meaning
$+ \cdot + = +$	A positive number times a positive number is a positive number.
$+ \cdot - = -$	A positive number times a negative number is a negative number.
$- \cdot + = -$	A negative number times a positive number is a negative number.
$- \cdot - = +$	A negative number times a negative number is a positive number.

$$(7x - 3y)(2x + 5y) = 14x^2 + 35xy - 6xy - 15y^2$$

Figure 1.4. *Expanding binomials horizontally using F-O-I-L.*

$$
\begin{array}{r}
7x - 3y \\
\times \quad 2x + 5y \\
\hline
35xy - 15y^2 \\
14x^2 - 6xy \quad\quad \\
\hline
14x^2 + 29xy - 15y^2
\end{array}
$$

Figure 1.5. *Multiplying binomials vertically.*

number. If there is an odd number of negative quantities multiplied together, their product is a negative number.

Occasionally, you may need to expand multiplication between the contents of two parentheses, such as $(7x - 3y)(2x + 5y)$. Expansion of this multiplication involves, as you may recall, *F-O-I-L (first-outer-inner-last)*, or a pattern of multiplying the first two terms, the outer two terms, the inner two terms, and the last two terms. See Figure 1.4. Finally, we combine like terms, $35xy$ and $-6xy$, to recognize the final expansion of the binomial multiplication to be $14x^2 + 29xy - 15y^2$. Alternatively, we can expand the two binomials vertically, just like we conduct regular whole number multiplication. See Figure 1.5. Either method of expanding binomial multiplication as shown in Figure 1.4 or Figure 1.5 is acceptable.

Where parentheses occur within parentheses, begin by simplifying the inner most set first. For example, $2(5 + (3 + 1)^2) = 2(5 + 4^2) = 2(5 + 16) = 2(21) = 42$.

1.2.2 Executing Exponents

Exponents are superscripts that tally the number of times a base is multiplied. For example, 4^3 is $4 \cdot 4 \cdot 4 = 64$, but $2 \cdot 4^3$ is $2 \cdot 4 \cdot 4 \cdot 4 = 2 \cdot 64 = 128$. In applying the exponent, care must be taken to apply it only to its immediate base, and not to any preceding multiplier. If an expression combines use of both parentheses and exponents, operations within parentheses are activated first, followed by operation of the exponent: for example, $(2 + 3)^2 = 5^2 = 25$ or $(2 \cdot 4)^3 = 8^3 = 512$. The expression $-3^2 = -(3^2) = -3 \cdot 3 = -9$ because the negative is not part of the base, while $(-3)^2 = (-3) \cdot (-3) = +9$ because parentheses are used to clearly designate the base of the exponent as -3. In an algebraic expression, the same order is used to expand $fm^2 = f \cdot m \cdot m$ but $(fm)^2 = f^2 \cdot m^2$. If the base of an exponent is an algebraic binomial, such as $(3x - 5y)^2$, expansion of the binomial expression follows the example shown in Figure 1.4 or Figure 1.5: $(3x - 5y) \cdot (3x - 5y) = 9x^2 - 30xy + 25y^2$.

Arithmetic operations can be conducted on exponents themselves. Exponents can be added, subtracted, multiplied, or divided as shown in Table 1.3.

Table 1.3. Rules for Operations with Exponents

Operation	Rule	Examples
Addition	Add exponents when the bases are the same and are multiplied together.	• $3^6 \cdot 3^4 = 3^{10}$ • $x^2 \cdot x^3 = x^5$
Subtraction	Subtract exponents when the bases are the same and are divided. Note: If the exponent in the denominator term is larger than the exponent in the numerator term, the subtraction produces a negative exponent, which simply means that the value is the reciprocal of the base raised to the positive exponent.	• $\dfrac{3^6}{3^4} = 3^2$ • $\dfrac{x^7}{x^3} = x^4$ • $\dfrac{x^3}{x^7} = x^{-4} = \dfrac{1}{x^4}$
Multiplication	Multiply exponents when a base raised to an exponent is itself raised to an exponent. Note: If the exponent is a fraction, multiply the exponents anyway. A base to the ½ exponent is the same as the square root of the base.	• $(4^3)^2 = 4^3 \cdot 4^3 = 4^6$ • $(2x^4)^3 = 2^3x^{12} = 8x^{12}$ • $\left(16x^5\right)^{\frac{1}{2}} = \left(4^2x^5\right)^{\frac{1}{2}}$ $= 4x^{\frac{5}{2}} = 4x^2x^{\frac{1}{2}} = 4x^2\sqrt{x}$
Division	Divide exponents by the root of the radical when a base raised to an exponent is contained in a radical.	• $\sqrt[3]{16x^7} = \sqrt[3]{2^4x^7}$ $= 2^{\frac{4}{3}}x^{\frac{7}{3}} = 2x^2 \cdot 2^{\frac{1}{3}}x^{\frac{1}{3}}$ or $= 2x^2 \cdot \sqrt[3]{2x}$

1.2.3   Multiply and Divide, then Add and Subtract

Continuing with the order of operations, we operate next with multiplication and division, whichever comes first left to right. So, $48 \div 4 \div 2 = 12 \div 2 = 6$. If you operate in the wrong order, you could end up with an incorrect answer of 24. After clearing parentheses, exponents, multiplication and division, we then add and subtract from left to right. Let's consider some examples as shown in Table 1.4.

Table 1.4. Summary Examples

Expression	Steps to Solution	Operation Used
1. $4 + 6 \cdot (4 + 1) \div 3 - 8 =$	$4 + 6 \cdot (4 + 1) \div 3 - 8 =$	Parentheses
	$4 + 6 \cdot 5 \div 3 - 8 =$	Multiplication
	$4 + 30 \div 3 - 8 =$	Division
	$4 + 10 - 8 =$	Addition
	$14 - 8 = 6$	Subtraction
2. $8 - 4 \div (5 - 3) \cdot 3 + 7$	$8 - 4 \div (5 - 3) \cdot 3 + 7 =$	Parentheses
	$8 - 4 \div 2 \cdot 3 + 7 =$	Division
	$8 - 2 \cdot 3 + 7 =$	Multiplication
	$8 - 6 + 7 =$	Subtraction
	$2 + 7 = 9$	Addition
3. $4 \cdot 8 + 12 \div 3 - 9 \cdot 4$	$4 \cdot 8 + 12 \div 3 - 9 \cdot 4$	Multiplication
	$32 + 12 \div 3 - 9 \cdot 4$	Division
	$32 + 4 - 9 \cdot 4$	Multiplication
	$32 + 4 - 36$	Addition
	$36 - 36 = 0$	Subtraction
4. $5 - 4 \cdot (2 + 1)^2 \div 6 + (2 + 3)^2$	$5 - 4 \cdot (2 + 1)^2 \div 6 + (2 + 3)^2 =$	Parentheses
	$5 - 4 \cdot 3^2 \div 6 + (2 + 3)^2 =$	Parentheses
	$5 - 4 \cdot 3^2 \div 6 + 5^2 =$	Exponent
	$5 - 4 \cdot 9 \div 6 + 5^2 =$	Exponent
	$5 - 4 \cdot 9 \div 6 + 25 =$	Multiplication
	$5 - 36 \div 6 + 25 =$	Division
	$5 - 6 + 25 =$	Subtraction
	$-1 + 25 = 24$	Addition
5. -4^2	$-4^2 = -16$	Exponent

1.3   Working with Equations

Being able to carry the standard order of operations into mathematical equations and formulas is an important next step. Below we introduce some frequently used equations from statistics and finance to expand our use and understanding of the order of operations. The reason for including the equations here is primarily to practice using the order of operations contained in them with sets of data, although a brief introduction of the equation and its terms are included. Each of the equations shown, then, is presented as an opportunity for practice, not as an object that the reader should necessarily be familiar with. The solutions are shown in detail so the reader should be able to follow the calculations, step by step. The discussions following each solution focus on the issues involved to arrive at the solution using the proper order of operations. In the next section of the chapter, we review the use of Excel in working with equations.

1.3.1   Statistical Equations

The field of statistics is rich with computational equations to use in summarizing and analyzing sets of data. Working with the summation function, denoted by an upper case sigma Σ, is a skill frequently employed. The summation function works as an individual unit and requires the valuation of the sum prior to performing any surrounding operations.

Example 1.1: Sample Mean

The mean is the most frequently used measure for the center of a set of data. To find a sample mean, denoted by the symbol $\overline{x}$, we sum the individual sample data values and divide by the number of observations sampled, using the equation:

$$\overline{x} = \frac{\sum_{i=1}^{n} x_i}{n}$$

where Σ means to sum the sampled values, x_i represents each of the individual values sampled, i is an index number indicating the position the value holds in the list of individual values, and n is the number of values sampled.

The Question

A sample of debt-to-equity ratios for 10 banks shows the following values: 5%, 2%, 7%, 4%, 3%, 6%, 3%, 4%, 3%, 9%. Find the mean debt-to-equity ratio for the sample.

Answer

$$\bar{x} = \frac{\sum_{i=1}^{10} x_i}{10} = \frac{5 + 2 + 7 + 4 + 3 + 6 + 3 + 4 + 3 + 9}{10} = \frac{46}{10} = 4.6$$

Using Excel

At the foot of the column or row containing the data, type: =average(range). The range represents the cells the data occupy on the spreadsheet. Alternatively, type: =sum(range), in another cell, type: =count(range), and then in a third cell, type: =(cell with sum)/(cell with count).

Discussion

The summation sign in the numerator is addressed first, because the entire numerator is divided by the denominator, 10. Once the numerator sum of 46 is obtained, we divide by 10 to find the sample mean of 4.6. The average debt-to-equity ratio for the sample is $\bar{x} = 4.6\%$.

Example 1.2: Variance

The variance is a frequently used measure that describes the concentration of data around the center of a data set. For a population, the variance is denoted by the symbol σ^2, pronounced *sigma squared*. The population variance is the sum of the squared differences of each value from its mean, μ, divided by N, the number of data values in the population. To be clear, the mean for an entire population is μ, whereas the mean for a sample of data selected from the population is $\bar{x}$.

$$\sigma^2 = \frac{\sum_{i=1}^{n}(x_i - \mu)^2}{N}$$

In contrast, the sample variance for a subset of data selected from a population is denoted by the symbol s^2. The sample variance is the sum of the squared differences of each value from its mean, $\bar{x}$, divided by $(n - 1)$, where the number of data values in the sample is n.

$$s^2 = \frac{\sum_{i=1}^{n}(x_i - \bar{x})^2}{n - 1}$$

The Question

There are five departments in a school of business. The numbers of full-time professors in each of the five departments are: 7, 10, 12, 9, and 5. Find the variance among the number of full-time professors in the school.

Answer

To find the population variance among the number of full-time professors in the school, we first have to find the population mean, μ, the average number of full-time professors across the five departments. The population mean is found by summing the five values and dividing by the number of departments in the school.

$$\mu = \frac{\sum_{i=1}^{5}x_i}{5} = \frac{7 + 10 + 12 + 9 + 5}{5} = \frac{43}{5} = 8.6$$

We then use the value of μ to compute the population variance, σ^2.

$$\sigma^2 = \frac{\sum_{i=1}^{n}(x_i - \mu)^2}{n}$$

$$= \frac{(7-8.6)^2 + (10-8.6)^2 + (12-8.6)^2 + (9-8.6)^2 + (5-8.6)^2}{5}$$

$$= \frac{2.56 + 1.96 + 11.56 + 0.16 + 12.96}{5} = \frac{29.2}{5} = 5.84$$

Using Excel

To compute the variance for the population, at the foot of the column or row containing the data, type: =varp(range). The range represents the cells the data occupy on the spreadsheet.

Discussion

Because the data given represent the entire population of departments in the school of business, we use the equation to compute the population variance, σ^2. We compute the numerator first because, like the computation in Example 1.1, the entire numerator is divided by the denominator. To compute the numerator, we subtract $\mu = 8.6$ from each of the five values in the numerator. We then square each of those differences and add the squared differences together. Finally, we divide by the total number of departments, 5. The population variance for the number of full-time professors in the school of business is $\sigma^2 = 5.84$.

Example 1.3: Sample Standard Deviation

The standard deviation is the positive square root of variance. For a population, the standard deviation is σ, pronounced *sigma*, and for a sample, the standard deviation is s.

$$s^2 = \frac{\sum_{i=1}^{n}(x_i - \bar{x})^2}{n-1}$$

Where the variance is given in squared units, the standard deviation is given in the same units the mean is reported in.

The Question

The number of knots in a sample of six boards of lumber is found to be: 1, 3, 1, 2, 0, and 2. Find the value of the sample standard deviation.

Answer

To find the sample standard deviation among the number of knots reported, we first have to find the average number of knots for the sample of boards, denoted by the symbol $\bar{x}$, using the equation:

$$\bar{x} = \frac{\sum\limits_{i=1}^{6} x_i}{6} = \frac{1+3+1+2+0+2}{6} = \frac{9}{6} = 1.5$$

The sample standard deviation is denoted by the symbol s. To calculate the sample standard deviation, we first square the differences between each value and the mean, sum the squared differences, divide by one less than the number of boards sampled, then take the square root of our answer. When the population mean, μ, is not known but is estimated by $\bar{x}$, the numerator of the calculation is divided by the sample size minus one, $(n-1)$. By dividing by one less than the sample size, we allow for more fluctuation in small samples to recognize potential error in using $\bar{x}$ to estimate μ.

$$
\begin{aligned}
s &= \sqrt{\frac{\sum\limits_{i=1}^{n}(x_i - \bar{x})^2}{n-1}} \\
&= \sqrt{\frac{(1-1.5)^2 + (3-1.5)^2 + (1-1.5)^2 + (2-1.5)^2 + (0-1.5)^2 + (2-1.5)^2}{6-1}} \\
&= \sqrt{\frac{0.25 + 2.25 + 0.25 + 0.25 + 2.25 + 0.25}{5}} = \sqrt{\frac{5.5}{5}} = \sqrt{1.1} = 1.049
\end{aligned}
$$

Using Excel

At the foot of the column or row containing the data, type: =stdev(range). The range represents the cells that are occupied by the data on the spreadsheet.

Discussion

We compute the numerator first because the entire numerator is divided by the denominator. To compute the numerator, we take each of the six values and subtract the sample mean of 1.5 from each. We then square each of those differences and add the squared differences together. We divide by one less than the number of boards inspected, 6 − 1 or 5. Finally, we take the square root of our answer. The sample standard deviation is approximately $s = 1.049$ knots for the sample of boards of lumber inspected.

Example 1.4: Estimated Sample Mean for Grouped Data

Sometimes managers may receive reports of data that have already been summarized into a frequency distribution. Alternatively, results of a study may be summarized in a publication in a way that does not include the mean. If the calculated mean is not included in the report, being able to back out an estimated mean is quite useful. For estimating either the population or the sample mean from grouped data, we use the concept of a weighted average by summing the product of the number of values in each class times the midpoint of its class as an approximation for the sum of all individual values sampled. We estimate the average miles per gallon for a sample using the equation:

$$\overline{x} = \frac{\sum_{i=1}^{k} f_i \cdot m_i}{n}$$

where $\overline{x}$ is the estimated sample mean, f_i is the frequency for each class i, m_i is the midpoint for each class i, k is the number of class intervals included, and n is the number of elements sampled. The same general equation is used to estimate the population mean, μ.

The Question

Shown in Table 1.5 and Figure 1.6 are data for the miles per gallon (MPG) ratings for city driving for 67 subcompact cars for the model year 2011. Estimate the average mileage for this sample of cars.

Answer

$$\bar{x} = \frac{\sum_{i=1}^{6} f_i \cdot m_i}{n} = \frac{7 \cdot 12 + 30 \cdot 16 + 23 \cdot 20 + 2 \cdot 24 + 5 \cdot 28 + 0 \cdot 32}{67}$$

$$= \frac{1212}{67} = 18.09$$

Using Excel

Arrange the frequency for each interval in one column and the midpoint of each interval in the adjacent column, aligned so that each interval frequency is in the same row as its midpoint. In a third column, multiply each interval's frequency times its midpoint. Excel uses the symbol * between two values to indicate multiplication, so the products will be formed by: =(cell with frequency)*(cell with midpoint). In Table 1.6, the products in cells D2 through D7 were formed with the equations shown in cells E2 through E7. Sum the frequencies to find the value of

Table 1.5. *City Driving Mileages, 2011 Model US Subcompact Cars*[1]

	A	B	C
	Class	Frequency	Midpoint
1			
2	10–14 MPG	7	12
3	14–18 MPG	30	16
4	18–22 MPG	23	20
5	22–26 MPG	2	24
6	26–30 MPG	5	28
7	30–34 MPG	0	32

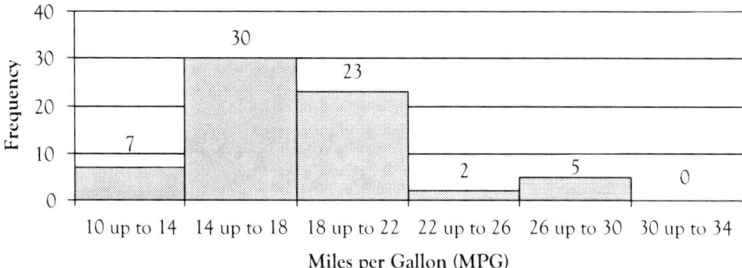

Figure 1.6. City Driving Mileages, 2011 Model US Subcompact Cars.[2]

Table 1.6. Using Excel to Calculate the Estimated Sample Mean[3]

	A	B	C	D	E
1	Class	Frequency	Midpoint	Product	Equations
2	10–14 MPG	7	12	84	=B2*C2
3	14–18 MPG	30	16	480	=B3*C3
4	18–22 MPG	23	20	460	=B4*C4
5	22–26 MPG	2	24	48	=B5*C5
6	26–30 MPG	5	28	140	=B6*C6
7	30–34 MPG	0	32	0	=B7*C7
8	SUMS:	67		1212	=sum(D2:D7)
9					
10	Estimated Sample Mean =		18.089552		=D8/B8

n by typing: =sum(range), as shown in cell B8. Sum the products created by multiplying each interval's frequency times its midpoint by typing: =sum(range), as shown in cell D8 with its equation shown in cell E8. In another cell, type: =(cell with the total of products)/(cell with sum of n), as calculated in cell C10 with its equation shown in cell E10.

Discussion

We compute the numerator first because the entire numerator is divided by the denominator. To compute the numerator, we multiply the frequency times the midpoint for each of the classes reported, sum across the products, and divide by the total number of elements sampled. The sample mean, $\bar{x}$, is estimated to be 18.09 miles per gallon for city driving for the sample of cars reported.

Example 1.5: Estimated Sample Standard Deviation for Grouped Data

To estimate the sample standard deviation from grouped data, we use the class frequencies, class midpoints, and the estimated sample mean in the equation:

$$s = \sqrt{\frac{\sum_{i=1}^{k} f_i \cdot m_i^2 - n \cdot \overline{x}^2}{n-1}}$$

where s is the estimated sample standard deviation, f_i is the frequency for each class i, m_i is the midpoint for each class i, k is the number of class intervals included, and n is the number of elements included.

The Question

Compute the estimated sample standard deviation for the data presented in Example 1.4.

Answer

To estimate the sample standard deviation for the miles per gallon (MPG) ratings for city driving for the 67 US subcompact cars, model year 2011 included in Example 1.4, we conduct the following calculations:

$$
\begin{aligned}
s &= \sqrt{\frac{\sum_{i=1}^{k} f_i \cdot m_i^2 - n \cdot \overline{x}^2}{n-1}} \\
&= \sqrt{\frac{(7 \cdot 12^2 + 30 \cdot 16^2 + 23 \cdot 20^2 + 2 \cdot 24^2 + 5 \cdot 28^2) - 67 \cdot 18.09^2}{67-1}} \\
&= \sqrt{\frac{22960 - 21926.62}{66}} \\
&= \sqrt{15.6724} = 3.96
\end{aligned}
$$

Using Excel

Arrange the frequency for each interval in one column and the midpoint of each interval in an adjacent column. In a third column, multiply each

interval's frequency times the square of its midpoint. Create a fourth column that contains the product of the frequency times the midpoint in
cells D2 through D7. Then create a fifth column in E2 through E7 by
multiplying the values in D2 through D7 by another factor of the midpoint, as shown in cells E2 through E7 in Table 1.7. In cell B8, sum the
frequencies to find the value of n by typing: =sum(B2:B7). In cell D8,
sum the frequencies in cells D2 through D7 by typing: =sum(D2:D7).
Compute the mean in cell B10 by typing: =D8/B8. In cell E8, sum
the products created by multiplying each interval's frequency times the
square of its midpoint by typing: =sum(E2:E7). That is the first component of the numerator to compute the sample variance. In cell E10, type:
=B8*(B10)^2. That forms the second component of the numerator to
compute the sample variance. In cell C12, type: =(E8−E10)/(B8−1). That
creates the value of the estimated sample variance in cell C12. Finally in
cell D13, type: =SQRT(C12), to form the estimated sample standard
deviation in cell D13.

Discussion

We compute the summation of fm^2 first, noting that the base for the
exponent 2 is only the midpoint value, m, for each class; the base of
the exponent does *not* include the class frequency, f. Also note that the
right-hand term in the numerator, $n \cdot \overline{x}^2$, is not a part of the summation
because the term is preceded by the subtraction sign. We first square the
midpoint values, then multiply the squared value times the frequency in
each class, and add those products across the classes. Secondly, we recognize the base for the exponent 2 in the right-hand term of the numerator
is only the sample mean, $\overline{x}$. So we square the sample mean and then
multiply that square by the 67 cars included in the report. We subtract
the two compound values in the numerator, subtract 1 from the number of cars to form the denominator of 66, and then divide. Finally we
take the square root of our answer. The estimated sample standard deviation for the city driving mileage for the 67 cars included in the report is
3.96 miles per gallon.

Table 1.7. Using Excel to Calculate the Estimated Sample Standard Deviation[4]

	A	B	C	D	E	F
1	Class	Frequency	Midpoint	f * m	f * m^2	Equations
2	10–14 MPG	7	12	84	1008	=B2*C2^2
3	14–18 MPG	30	16	480	7680	=B3*C3^2
4	18–22 MPG	23	20	460	9200	=B4*C4^2
5	22–26 MPG	2	24	48	1152	=B5*C5^2
6	26–30 MPG	5	28	140	3920	=B6*C6^2
7	30–34 MPG	0	32	0	0	=B7*C7^2
8	SUMS:	67		1212	22960	=sum(E2:E7)
9						
10	x-bar =	18.089552		n*(x-bar)^2 =	21924.53673	=B8*B10^2
11						
12	Estimated Sample Variance =		15.688837			=(E8-E10)/(B8-1)
13	Estimated Sample Standard Deviation =		3.96091			=SQRT(C12)

Example 1.6: The z-Score for a Sample Mean

The z-score is the scale on the axis along the bottom of the standard normal distribution. It represents the number of units of standard deviation a particular sample mean is above or below the population mean. The standard normal distribution is useful because its table details the amount of area captured under the normal curve below a given value. That area also represents the likelihood, or probability, that a value will fall within a defined segment of a normal distribution. The z-score is calculated using the equation:

$$z = \frac{\bar{x} - \mu}{\sigma / \sqrt{n}}$$

where $\bar{x}$ is the sample mean, μ is the population mean, σ is the population standard deviation, and n is the sample size.

The Question

Annual consumption of chicken in the United States is on the rise. Increasingly, it's what's for dinner. Population census figures place the average annual consumption of chicken at 90.6 lbs per person.[5] Assuming the population standard deviation of 10 lbs in the annual consumption of chicken per capita in the United States, what is the z-score for a sample mean of 93.2 lbs of chicken consumed per person last year for a sample of 45 people in the United States?

Answer

$$z = \frac{(\bar{x} - \mu)}{\sigma / \sqrt{n}} = \frac{(93.2 - 90.6)}{10 / \sqrt{45}} = \frac{(93.2 - 90.6)}{10 / 6.708} = \frac{2.6}{1.4907} = 1.74$$

Using Excel

In a cell, type: =(value for sample mean—value for population mean)/(value for population standard deviation/SQRT(n)).

Discussion

We begin simplifying the denominator, evaluating the square root of 45, then dividing 10 by that value. We simplify the numerator and finally divide the numerator by the denominator. The resulting z-score rounds to 1.74.

Example 1.7: Pooled Variance Estimate, Two Populations

In statistics, we use a statistical test to settle the question of whether two population means are different. Sometimes we conduct that test assuming their two variances are roughly equal. When we assume the two variances are equal, we act accordingly and combine the two sample variances to form a single pooled variance estimate, s_p^2. The pooled variance estimate is calculated from the two sample variances weighted by one less than their respective sample sizes and averaged across their combined sample sizes minus two. The equation we use to combine the two sample variances is:

$$s_p^2 = \frac{(n_1 - 1) \cdot s_1^2 + (n_2 - 1) \cdot s_2^2}{n_1 + n_2 - 2}$$

where s_p^2 is the pooled variance estimate, n_1 is the number sampled from population 1, s_1^2 is the sample variance among the values sampled from population 1, n_2 is the number sampled from population 2, and s_2^2 is the sample variance among the values sampled from population 2.

The Question

A diversified company set up separate e-commerce sites to handle orders for two of the company's product lines. Internal auditors randomly selected 32 one-hour periods when they recorded the number of orders placed on site A and 36 periods when they recorded the number of orders placed on site B. Find the pooled variance estimate for the number of orders placed on the two sites if the variance for the first sample is 7.8^2 and for the second sample is 9.7^2.

Answer

$$s_p^2 = \frac{(n_1 - 1) \cdot s_1^2 + (n_2 - 1) \cdot s_2^2}{n_1 + n_2 - 2} = \frac{(32 - 1) \cdot 7.8^2 + (36 - 1) \cdot 9.7^2}{32 + 36 - 2}$$

$$= \frac{31 \cdot 60.84 + 35 \cdot 94.09}{32 + 36 - 2} = \frac{1886.04 + 3293.15}{32 + 36 - 2} = \frac{5179.19}{66} = 78.47$$

Using Excel

Assuming the individual values for each of the two samples are available, at the foot of the two columns or rows containing the sample data, type: =count(range) for each sample. That generates the sample size for each sample. In nearby cells, type: =var(range) for each sample. In another cell, type: =(([cell with count for sample 1]−1)*[cell with variance for sample 1]+([cell with count for sample 2]−1)*[cell with variance for sample 2])/([cell with count for sample 1]+[cell with count for sample 2]−2). We will discuss some of the automated functions built into Excel's Data Analysis Toolkit later in Chapter 3.

Discussion

We clear parentheses in the numerator, getting 31 and 35 on each term in the numerator, respectively, then square each of the two separate standard deviations. We multiply the values in each of those two terms then add the products together to form the numerator. We simplify the denominator using addition then subtraction. Finally we divide the numerator by the denominator. The pooled variance estimate for the two samples taken is 78.47.

Example 1.8: *t*-Score for Two Means from Populations with Equal Variances

The rationale for this example involves some complex statistical notation. The reader can skip this general introduction and still be able to evaluate the equation in terms of the order of operations performed.

The general introduction revolves around a test of two means. When we want to settle the question of whether two populations have different means, we conduct a statistical test that measures how far from zero the difference of their two means is in terms of their common standard error. If we do not know the population variances, but they are roughly equivalent, and some minimum conditions hold for the two populations, we use the two sample variances as a basis for the value of their combined standard error and evaluate a t-score using the following equation:

$$t = \frac{\left(\bar{x}_1 - \bar{x}_2\right) - \left(\mu_1 - \mu_2\right)}{\sqrt{s_p^2 \cdot \left(\dfrac{1}{n_1} + \dfrac{1}{n_2}\right)}}$$

where $\bar{x}_1$ and $\bar{x}_2$ are the means from the two samples taken from populations 1 and 2, $(\mu_1 - \mu_2)$ is zero because we are settling the question of whether the difference between the two means is different from zero, s_p^2 is the pooled variance estimate shown earlier in Example 1.7, and n_1 and n_2 are the number of elements selected for each of the two samples. Technically, when we are settling the question of whether the difference between the two means is different from zero, we do not have to include that second term in the numerator, $(\mu_1 - \mu_2)$, because it is assumed to equal zero.

The Question

A diversified company set up separate e-commerce sites to handle orders for two of the company's product lines. Internal auditors randomly selected 32 one-hour periods when they recorded the number of orders placed on site A and 36 periods when they recorded the number of orders placed on site B. The variance for the first sample is 7.8^2 and for the second sample is 9.7^2, and their respective sample means are 31.75 and 37. Use the pooled variance estimate of 78.47 found in Example 1.7.

Answer

$$t = \frac{\left(\bar{x}_1 - \bar{x}_2\right) - \left(\mu_1 - \mu_2\right)}{\sqrt{s_p^2 \cdot \left(\dfrac{1}{n_1} + \dfrac{1}{n_2}\right)}} = \frac{(31.25 - 37) - 0}{\sqrt{78.47 \cdot \left(\dfrac{1}{32} + \dfrac{1}{36}\right)}}$$

$$= \frac{-5.25}{\sqrt{78.47 \cdot (0.059)}} = \frac{-5.25}{\sqrt{4.6319}} = \frac{-5.25}{2.152} = -2.439$$

Using Excel

Compute the pooled variance estimate as detailed in Example 1.7. Compute the sample means as detailed in Example 1.1. Here we assume we are looking to evaluate whether there is any difference between the two population means, which means the value of $(\mu_1 - \mu_2)$ is zero. In a separate cell, type: =(cell with mean sample 1 − cell with mean sample 2)/ SQRT(value of pooled variance estimate*(1/cell with count sample 1+ 1/cell with count sample 2)).

Discussion

Subtract the two means in the numerator. To evaluate the denominator, we start inside the parentheses by adding the two fractions. We multiply the sum of the two fractions times the value of the pooled variance estimate, then take the square root of the product. Finally, we divide the numerator by the denominator. The t-score for the difference between the two means is $t = -2.439$.

1.3.2 Financial Equations

While use of the summation function is prominent in many of the calculations conducted with statistical equations, the use of exponents is important in working with financial equations.

Example 1.9: Compound Interest, Future Value

To capture the effect of a lump sum of money left in an account to compound over a number of interest periods, we use the equation:

$$A = P(1 + i)^n$$

where A is the future value of P dollars deposited in an account that earns $i\%$ interest in each of the n interest periods. Interest rates are often quoted on an annual basis as $r\%$ rather than the rate allocated per interest period, $i\%$. When the interest is computed monthly, $i\% = \dfrac{r\%}{12}$. When the interest is computed quarterly, $i\% = \dfrac{r\%}{4}$. The time over which money accrues interest is more often quoted in years than the number of interest periods the account is active. Some adjustments may be necessary to fit the input to the equation. When the interest is computed monthly, then $n = t \cdot 12$. When the interest is computed quarterly, then $n = t \cdot 4$.

The Question

An initial deposit of $1,200 is made into an account earning 3.25% interest compounded quarterly. If the account is left to accrue interest, how much will be in the account after 5 years?

Answer

$$r = 0.0325, \text{ so } i = \frac{0.0325}{4} = 0.008125$$

$$t = 5 \text{ years, so } n = 5 \cdot 4 = 20 \text{ interest periods}$$

$$A = 1200 \cdot (1 + 0.008125)^{20} = 1200 \cdot 1.175676 = 1410.81$$

Using Excel

In a cell, type: =1200*(1+0.008125)^20.

Discussion

We adjust the annual interest rate, r, to a periodic rate, i, by dividing r by 4, and we adjust the number of years, t, to the number of interest periods, n, over which the account is active by multiplying t by 4. We convert i from a percent to a decimal, add 1, and raise the sum to the nth power. That interim value, in this example $1.175676, is the future value of one dollar left in the account at $i = 0.008$ periodic interest rate for $n = 20$ interest periods. Since we began the account with $1,200, we finally multiply the initial principal times the future value of a dollar to arrive at a future value of $1,410.81 for the account.

Example 1.10: Compound Interest, Present Value

If we conduct some simple algebra on the general equation contained in Example 1.9, we can solve the equation to give us what lump sum we must deposit in an account in order to have a specified value at some future date. Dividing both sides by $(1+i)^n$, we derive the equation:

$$P = \frac{A}{(1+i)^n} \quad \text{or} \quad A(1+i)^{-n}$$

The Question

Suppose Maria has a balloon payment of $7,450 in 3 years on the new car she just leased. Unbeknownst to her, her father decides to open an account with a deposit that will cover the balloon payment at the end of her lease. If the account earns 3% compounded monthly, how much will he have to deposit in the account now to have enough to cover her balloon payment at the end of her 3-year lease?

Answer

$$r = 0.03, \text{ so } i = \frac{0.03}{12} = 0.0025$$

$t = 3$ years, so $n = 3 \cdot 12 = 36$ interest periods

$$P = \frac{7450}{(1+0.0025)^{36}} = \frac{7450}{1.094051} = 6809.55$$

or

$$P = 7450 \cdot (1+0.0025)^{-36} = 7450 \cdot 0.914034 = 6809.55$$

Using Excel

In a cell, type: =7450(1+0.0025)^−36.

Discussion

The future value of a dollar invested today in an account earning 3% interest compounded monthly across 3 years is $1.094051. So we divide

$7,450 by the future value to find out how many dollars we need to invest today to accrue a total of $7,450 in 3 years. Alternatively, the present value of one dollar 3 years from now is $0.914034, meaning we have to invest slightly more than $0.91 today for every dollar we want to have 3 years from now. We can multiply $7,450 times today's value of a future dollar to determine how much we have to invest today to have $7,450 in 3 years. You will notice that

$$\frac{1}{1.094051} = (1.094051)^{-1} = 0.914034$$

The present value of $7,450 3 years from now under 3% compounded monthly is $6,809.55, the amount her father will need to invest today to have the full amount of Maria's balloon payment in 3 years.

Example 1.11: Effective Interest Rate

The effective interest rate is the simple interest rate that, applied to a compounding account, would yield the same interest.

$$r_e = \left(1 + \frac{r}{m}\right)^m - 1$$

where r_e is the effective interest rate, r is the stated rate of interest compounding m times in a year.

The Question

Credit card accounts are an excellent environment to understand effective rates. The stated rate of interest on a credit card is 18%, but the rate is compounded every month. What is the effective interest rate?

Answer

$$r_e = \left(1 + \frac{r}{m}\right)^m - 1 = \left(1 + \frac{0.18}{12}\right)^{12} - 1 = 0.1956$$

Using Excel

In a cell, type: =(1+.18/12)^12−1. In the next chapter, we review an auto-mated function in Excel to calculate the effective interest rate.

Discussion

The equation is straightforward to compute. Inside the parentheses, we compute the periodic interest rate by dividing r by m. We then raise the sum of 1 plus the periodic interest rate to the number of interest periods in the year, and finally subtract 1. While the credit card company may advertise 18% interest on the unpaid balance, the effective interest rate is much higher, at 19.56%.

Example 1.12: Compound Interest, Future Value of an Ordinary Annuity

Examples 1.9 and 1.10 dealt with accounts that were opened with a single deposit into the account. In contrast, this example deals with a regular stream of payments made into a savings account, sometimes referred to as *an annuity* or *a sinking fund*. If the payments are made at the end of the interest period, it is called an *ordinary annuity*. The future value of a regular stream of payments into the account can be calculated using the equation:

$$S = R \cdot \left(\frac{(1+i)^n - 1}{i} \right)$$

where R = the amount of the payment, i = the periodic interest rate, and n = the number of payments made.

The Question

To celebrate the birth of their daughter, Rich and Diane decided to open an account with a deposit of $25 at the end of the first month. They set up an automatic deposit so that at the end of every month, $25 is deposited into the account. The account earns 3% compounded monthly. How much will be in the account at the end of the month in which their daughter turns 21?

Answer

Adjusting the interest rate to a monthly figure yields $i = 0.03/12 = 0.0025$ and the time of 21 years to months yields $n = 21 \cdot 12 = 252$. So, over the years, Rich and Diane make 252 deposits of $25 each; they have put $252 \cdot \$25$ or $6,300 in principal into the account. But the account earns interest that compounds each month. Using the equation for the future value of an annuity, we find the following.

$$S = R \cdot \left(\frac{(1+i)^n - 1}{i} \right)$$

$$= 25 \cdot \left(\frac{(1+0.0025)^{252} - 1}{0.0025} \right) = 25 \cdot \left(\frac{1.876135 - 1}{0.0025} \right) = 25 \cdot 350.454$$

$$= 8761.35$$

Using Excel

In a cell, type: =FV(.0025,252,–25,0,0). Alternatively, type: =25*(((1+.0025)^252–1)/.0025). In the next chapter, we review an automated function in Excel to calculate the future value of an annuity.

Discussion

Each dollar committed to a stream of 252 monthly payments earning 3% compounded monthly generates $350.454 because of the interest earned over the intervening months. Although the interest rate of 3% is low, each dollar committed is worth much more because of the number of periods over which interest compounds, nearly 40% more:

$$\frac{350.454}{252} = 1.39069$$

The account will be worth $8,761.35 at the end of the month in which their daughter turns 21.

Example 1.13: Present Value of an Ordinary Annuity

The present value of a stream of regular payments is appropriate when a consumer purchases an asset and pays for it over time. If the consumer takes possession of an asset now and is paying it off into the future, the lender gets the benefit of the interest, not the individual who is making the purchase. So the individual making the purchase pays the regular principal due plus interest. The present value of a regular stream of payments can be calculated using the equation:

$$P = R \cdot \left(\frac{1 - (1+i)^{-n}}{i} \right)$$

where R = the amount of the payment, i = the periodic interest rate, and n = the number of payments made.

The Question

Harry decides to add a new delivery truck to his fleet. With tax and license, the truck costs \$24,681.30. Harry puts \$2,681.30 down and carries the rest of the cost, \$22,000, on a loan that is scheduled to pay off in 3 years for 3.6% interest compounded monthly. What are his monthly payments?

Answer

Adjusting the interest rate to a monthly figure yields i = 0.036/12 = 0.003 and the time of 3 years to months yields n = 3 · 12 = 36. Without calculating the cost of the interest, Harry owes \$22,000 which he will pay back in 36 payments of \$611.11 per payment. But Harry owes interest each month as well. To use the equation for the present value of an annuity, we need to solve for R, the monthly payment, given we know P, the amount of the loan he took out:

$$R = \frac{P}{\left(\dfrac{1 - (1+i)^{-n}}{i} \right)} = \frac{22,000}{\left(\dfrac{1 - (1+0.003)^{-36}}{0.003} \right)} = \frac{22,000}{\left(\dfrac{1 - 0.897773}{0.003} \right)}$$

$$= \frac{22,000}{34.0757554} = 645.6203$$

Using Excel

In a cell, type: =PMT(.003,36,−22000,0,0). Alternatively, in a cell, type: =22000*((1−(1+.003)^−36)/.003). In the next chapter, we review an automated function in Excel to calculate the present value of an annuity.

Discussion

The value of Harry's monthly payment for the loan he negotiated on his truck purchase is $645.62 per month. Over the life of the loan, Harry will pay a total of $645.62 × 36 = $23,242.32, which includes ($23,242.32 − $22,000) = $1,242.32 to cover the interest due through the life of the loan.

Example 1.14: Internal Rate of Return

The internal rate of return represents the annual rate of appreciation in value from funds invested in a multi-year project. Where the other finance equations included in this chapter rely on equal payments, the internal rate of return allows for differing values to be recognized over a project's life span. Calculations for the internal rate of return require that the same period of time be reflected in all expected costs and anticipated incomes. All other factors being equal, the project with the highest internal rate of return is comparatively preferable to projects with lower internal rates of return. The programmed Excel function =IRR(range,estimate for the rate) reflects the comparative ease with which Excel simplifies the computations.

The Question

Suppose a company is considering launching a project which they estimate will initially cost $50,000 the first year. Years 2 through 6 are projected to see revenues of $17,000; $25,000; $30,000; $32,000; and $35,000. What is the anticipated internal rate of return?

Answer

Using Excel

In an Excel spreadsheet, enter the values –50000, 17000, 25000, 30000, 32000, 35000. In a cell, type: =IRR(range,.1). Excel returns the answer to the complex calculations as 41%. Our guess of 0.1 or 10% was very low, but adequate for Excel to use as a beginning approximation of the internal rate of return.

Discussion

The internal rate of return is 41% and represents the interest rate accruing to a project that reflects both start-up costs and anticipated incomes.

Example 1.15: Net Present Value

The net present value of an investment is the present value of anticipated cash payments and cash incomes associated with a project discounted to reflect value in today's dollars. Where the internal rate of return reports project activity as a rate of anticipated growth, the net present value reports a dollar value of the investment, the value in today's dollar that, deposited in a bank to accrue at the projected rate, would equal the difference of payments and incomes recognized over equal periods of time.

The Question

Suppose a company is considering launching a project which they estimate will initially cost $50,000 the first year. Years 2 through 6 are projected to see revenues of $17,000; $25,000; $30,000; $32,000; and $35,000. Given an 8% discount rate, what is the net present value of this stream of activity?

Answer

Using Excel

In an Excel spreadsheet, enter the values −50000, 17000, 25000, 30000, 32000, 35000. In a cell, type: =NPV(.08,range). Excel returns the answer to the calculations as $54,009.76.

Discussion

The current value of the anticipated payments and incomes is $54,009.76, the amount one could place on deposit today earning 8% compounded annually, which at the end of 6 years, would equal the effect of the project over the next 6 years.

CHAPTER 2

Working with Automated Functions in Excel

Microsoft Excel has some powerful quantitative tools built into its capabilities. This chapter will continue the introduction of those that are captured in the Insert Function, mentioned informally in the examples captured in the latter half of Chapter 1. Located near the formula bar in Microsoft Excel is an Insert Function toolbar button: $\boxed{fx}$. It may appear in different locations on different computer displays, but it works in the same way across different versions of Excel. The button activates a list of categories and, within each category, a sublist of functions. In this chapter, we investigate some of the functions listed under the Statistical, Math & Trig, and Financial categories.

2.1   Inserting Selected Statistical Functions

The Statistical category under the Insert Function toolbar button includes a list of nearly 80 functions. We introduce the basic descriptive statistical functions here.

2.1.1   AVERAGE

The arithmetic mean, or average, is the sum of the individual data values divided by the number of observations. It is presented in Chapter 1, Example 1.1 and is the most frequently used measure for the center of a set of data.

To find the average for a set of data, locate your cursor in a spreadsheet cell, activate the Insert Function toolbar button, and select the category: Statistical. In the large window, select the function: AVERAGE. With your cursor active in the field for Number 1, scroll over the data for which you want to calculate the mean on the spreadsheet beneath the window, and click OK. The value for the average will appear in the spreadsheet cell. Alternatively, the arithmetic mean for a set of data can be found directly in Excel. Simply type the function: =average(range of data) in a spreadsheet cell.

Example 2.1: Finding the Average

The average, or mean, is the most frequently used measure for the center of a set of data. The mean, denoted for a sample by the symbol $\bar{x}$, represents the sum of the individual data values divided by the number of observations sampled.

The Question

A sample of eight different vehicles in the corporate car pool is selected and their city mileages per gallon of fuel are measured, producing the values 19.7, 17.6, 21.4, 18.3, 19.5, 18.2, 19.0, and 18.9 mpg. Find the average mileage for the sample.

Answer

$\bar{x} = 19.075$

Using Excel

Enter the data in individual cells A1 through A8. With your cursor in cell A9, activate the Insert Function button, select the Statistical category, highlight average, and click OK. Activate your cursor in the field for Number 1 and then scroll your cursor over cells A1 through A8 containing the data. Click OK. Excel will return the value of the average, 19.075, in cell A9. This is equivalent to the procedure outlined in Chapter 1, Example 1.1.

Discussion

The average mileage for the sample of eight cars is $\bar{x} = 19.075$ mpg. Across the eight cars, on average they travel 19.075 miles for each gallon of fuel.

2.1.2   COUNT

The count function is valuable in determining how many numeric elements are in a column or row of data. The count function will not include text values or empty cells in its total, although cells containing a "0" will be included in the total. To find the count for a set of data, locate your cursor in a spreadsheet cell, activate the Insert Function toolbar button, and select the category: Statistical. In the large window, scroll down the list and select the function: COUNT. With your cursor active in the field for Value 1, scroll over the data for which you want to determine the count on the spreadsheet beneath the window, and click OK. The value for the count of numeric items in that range will appear in the spreadsheet cell. Alternatively, the count for a set of data can be found directly in Excel. Simply type the function: =count(range of data) in a spreadsheet cell.

Example 2.2: Counting Data

To practice the process of counting data elements, we will return to the data for Example 2.1 that were entered in cells A1 through A8 on the Excel spreadsheet.

The Question

A sample of eight different vehicles in the corporate car pool is selected and their city mileages per gallon of fuel are measured, producing the values 19.7, 17.6, 21.4, 18.3, 19.5, 18.2, 19.0, and 18.9 mpg. Verify the number of mileage measurements in the sample.

Answer

$x = 8$

Using Excel

Enter the data in individual cells A1 through A8. With your cursor in cell A9, activate the Insert Function button, select the Statistical category, highlight COUNT, and click OK. Activate your cursor in the field for Value 1 and then scroll your cursor over cells A1 through A8 containing the data. Click OK. Excel will return the count of values, 8, in cell A9.

Discussion

There are eight measurements entered on the spreadsheet. When large data sets are processed, having exact counts may not be as obvious as our simple example shown here.

2.1.3   COUNTIF

The countif function is a versatile function that allows you to establish conditions and determine the total number of elements that meet the stated conditions. We consider the countif function as it operates with three different types of data below.

2.1.3.1   Numeric, Exact Values

If the data to be counted are integers, let's say they are whole numbers from 1 to 5, you can get a complete breakdown of the totals as follows.

1. Activate the Insert Function toolbar button, and select the category: Statistical. In the large window, scroll down the list and select the function: COUNTIF. With your cursor active in the field for Range, scroll over the data for which you want to determine the count on the spreadsheet beneath the window. In the field for Criterion, type the value 1, and click OK. Alternatively, type =countif(range of data,1) in a spreadsheet cell. The spreadsheet cell will display the number of elements in the range that are exactly equal to 1.

2. Repeat the process above, but type the value 2 in the Criterion field. The spreadsheet cell will display the number of elements in the range that are exactly equal to 2.

3. Repeat the process for each of the additional integer values over which you want a count.

2.1.3.2   Numeric, Intervals of Values

If the data to be counted are numeric but are noninteger values, let's say they are from 1 to 5, you can get a complete breakdown of the totals as follows.

1. Activate the Insert Function toolbar button, and select the category: Statistical. In the large window, scroll down the list and select the function: COUNTIF. With your cursor active in the field for Range, scroll over the data for which you want to determine the count on the spreadsheet beneath the window. In the field for Criterion, type the value 1, and click OK. Alternatively, type: =countif(range of data, 1) in a spreadsheet cell. The spreadsheet cell will display the number of elements in the range that are exactly equal to 1.

2. Repeat the process above, but type in the Criterion field: "<3". The spreadsheet cell will display the number of elements in the range that have a value less than 3. To identify the number of elements that are greater than 1 but less than 3, subtract the count of elements that are exactly equal to 1 from the total number of elements that are less than 3. Alternatively, type: =countif(range of data, "<3")— countif(range of data, 1) in a spreadsheet cell. The value in the cell represents the total number of values that are less than 3 but greater than 1.

3. Repeat the process above, but type in the Criterion field: "<4". The spreadsheet cell will display the number of elements in the range that have a value less than 4. To identify the number of elements that are greater than 1 but less than 4, subtract the count of elements that are less than 3 from the total number of elements that are less than 4. Alternatively, type: =countif(range of data, "<4")—countif(range of data, "<3") in a spreadsheet cell. The value in the cell represents the total number of values that are less than 4 but greater than 3.

4. To identify the number of elements that are exactly equal to 5, activate the Insert Function toolbar button, and select the category: Statistical. In the large window, scroll down the list and select the function: COUNTIF. With your cursor active in the field for Range, scroll over the data for which you want to determine the count on the spreadsheet beneath the window. In the field for Criterion, type

the value 5, and click OK. Alternatively, type: =countif(range of data, 5) in a spreadsheet cell. The spreadsheet cell will display the number of elements in the range that are exactly equal to 5.

2.1.3.3   Nonnumeric, Text Data

If the data are nonnumeric text data, let's say they are left, right, and center, you can get a complete breakdown of the totals as follows.

1. Activate the Insert Function toolbar button, and select the category: Statistical. In the large window, scroll down the list and select the function: COUNTIF. With your cursor active in the field for Range, scroll over the data for which you want to determine the count on the spreadsheet beneath the window. In the field for Criterion, type: "left", and click OK. Alternatively, type: =countif(range of data,"left") in a spreadsheet cell. The spreadsheet cell will display the number of elements in the range that are labeled "left".

2. Activate the Insert Function toolbar button, and select the category: Statistical. In the large window, scroll down the list and select the function: COUNTIF. With your cursor active in the field for Range, scroll over the data for which you want to determine the count on the spreadsheet beneath the window. In the field for Criterion, type: "center", and click OK. Alternatively, type: =countif(range of data, "center") in a spreadsheet cell. The spreadsheet cell will display the number of elements in the range that are labeled "center".

3. Activate the Insert Function toolbar button, and select the category: Statistical. In the large window, scroll down the list and select the function: COUNTIF. With your cursor active in the field for Range, scroll over the data for which you want to determine the count on the spreadsheet beneath the window. In the field for Criterion, type: "right", and click OK. Alternatively, type: =countif(range of data,"right") in a spreadsheet cell. The spreadsheet cell will display the number of elements in the range that are labeled "right".

The countif function applied to text data is not sensitive to capital versus lower case letters as long as the criterion specified in the quotations is itself lower case. The resulting count will include all items with the specified text string whether the items are capitalized or not. If, however, the criterion specified in the quotations is capitalized, the resulting count will only include those items that begin with a capital letter. In the event, a count of just those elements with the specified text string where the items are not capitalized is needed, an exact count can be achieved by conducting the countif function for the criterion specified with the first letter in lower case minus the countif function for the criterion specified with the first letter in upper case.

Example 2.3: Counting Using Criteria

To practice the process of counting data elements that meet stated criteria, we will return to the data for Example 2.1 that were entered in cells A1 through A8 on the Excel spreadsheet.

The Question

A sample of eight different vehicles in the corporate car pool is selected and their city mileages per gallon of fuel are measured, producing the values 19.7, 17.6, 21.4, 18.3, 19.5, 18.2, 19.0, and 18.9 mpg. Verify the number of mileage measurements in the sample that are greater than 19.

Answer

$x = 3$

Using Excel

Enter the data in individual cells A1 through A8. With your cursor in cell A9, activate the Insert Function button, select the Statistical category, highlight COUNTIF, and click OK. Activate your cursor in the field for Range and scroll your cursor over cells A1 through A8 containing the data. Next locate your cursor in the field for Criterion and type: ">19". Click OK. Excel will return the count of values that fall above 19, 3, in cell A9.

Discussion

There are three measurements entered on the spreadsheet that are greater than 19. Note that the value 19.0 is not included in the count, since it is not greater than itself. When large data sets are processed, having exact counts of data that meet stated criteria may be very useful.

2.1.4   MAX

Sometimes it is useful to know the maximum value in a set of data. Along with the minimum value of the data set, the maximum establishes a preliminary sense of the spread among data given by the range over which the data vary, from the smallest value to the largest value in the data set. The max function identifies the largest value in a set of numeric elements in a column or row of data. To find the maximum for a set of data, locate your cursor in a spreadsheet cell, activate the Insert Function toolbar button, and select the category: Statistical. In the large window, scroll down the list and select the function: MAX. With your cursor active in the field for Number 1, scroll over the data for which you want to determine the maximum on the spreadsheet beneath the window, and click OK. The maximum value of the numeric items in that range will appear in the spreadsheet cell. Alternatively, the maximum for a set of data can be found directly in Excel. Simply type the function: =max(range of data) in a spreadsheet cell.

The largest value across two or more columns or rows of data can also be found using the max function. With your cursor active in the field for Number 1, scroll over the data in the first column or row you want to include in your search for the maximum on the spreadsheet beneath the window. With your cursor active in the field for Number 2, scroll over the data in the second column or row you want to include in your search for the maximum on the spreadsheet beneath the window. You will notice that, as soon as you activate your cursor in the field for Number 2, the window automatically expands to allow for a third column or row to be included. The max function allows you to include up to 30 columns or rows of data in any single search. Alternatively, the maximum for a set of data in multiple columns or rows can be found directly in Excel. Simply type the function: =max(range of data Number 1,range for data Number 2, …) in a spreadsheet cell.

Example 2.4: Finding the Maximum Value

To find the maximum value of a set of data, we will return to the data for Example 2.1 that were entered in cells A1 through A8 on the Excel spreadsheet.

The Question

A sample of eight different vehicles in the corporate car pool is selected and their city mileages per gallon of fuel are measured, producing the values 19.7, 17.6, 21.4, 18.3, 19.5, 18.2, 19.0, and 18.9 mpg. Find the maximum value in the set of data.

Answer

$x = 21.4$

Using Excel

Enter the data in individual cells A1 through A8. With your cursor in cell A9, activate the Insert Function button, select the Statistical category, highlight MAX, and click OK. Activate your cursor in the field for Number 1 and scroll your cursor over cells A1 through A8 containing the data. Click OK. Excel will return the maximum value of the set, 21.4, in cell A9.

Discussion

No car registered more than 21.4 mpg in the sample of eight vehicles.

———————————

2.1.5   MEDIAN

The median is the value above which and below which half of the values of a set of data fall when the data are put into an ordered array. In general, to find a median:

- For an *odd* number of observations, the median is the middle number when the data are put in an ordered array.
- For an *even* number of observations, the median is the average of the middle two values when the data are put in an ordered array.

Finding the median is independent of whether the data are ordered from smallest to largest or largest to smallest. Unlike the average, the value of the median is not influenced by the presence of outliers and may provide a more reliable estimate of a distribution's central value when outliers are present. In discussions of residential housing values, for example, we frequently see references to median home values in lieu of average home values because of the potential bias introduced by a few high-value homes into the calculation of the mean home value in a given market.

The median for a set of numeric elements in a column or row of data can be identified easily using the median function. To find the median for a set of data, locate your cursor in a spreadsheet cell, activate the Insert Function toolbar button, and select the category: Statistical. In the large window, scroll down the list and select the function: MEDIAN. With your cursor active in the field for Number 1, scroll over the data for which you want to determine the median on the spreadsheet beneath the window, and click OK. The median value of the numeric items in that range will appear in the spreadsheet cell. Alternatively, the median for a set of data can be found directly in Excel. Simply type the function: =median(range of data) in a spreadsheet cell. As with the max function, the median for two or more columns or rows of data can also be found.

Example 2.5: Finding the Median

To find the median of a set of data, we will return to the data for Example 2.1 that were entered in cells A1 through A8 on the Excel spreadsheet.

The Question

A sample of eight different vehicles in the corporate car pool is selected and their city mileages per gallon of fuel are measured, producing the values 19.7, 17.6, 21.4, 18.3, 19.5, 18.2, 19.0, and 18.9 mpg. Find the median in the set of data.

Answer

$x = 18.95$

Using Excel

Enter the data in individual cells A1 through A8. With your cursor in cell A9, activate the Insert Function button, select the Statistical category, highlight MEDIAN, and click OK. Activate your cursor in the field for Number1 and scroll your cursor over cells A1 through A8 containing the data. Click OK. Excel will return the median of the set, 18.95, in cell A9.

Discussion

Since there are an even number of cars included the sample, the median will be the average of the middle two values once the values are sorted in ascending or descending order. When sorted in ascending order, 18.9 is the fourth value and 19.0 is the fifth value. Their average is 18.95. Likewise, when the values are sorted in descending order, 19.0 is the fourth value and 18.9 is the fifth value. Again their average is 18.95. Half of the values fall above 18.95 and half of the values fall below 18.95.

2.1.6   MIN

Sometimes it is useful to know the minimum value in a set of data. Together with the maximum value of the data set, the minimum establishes a preliminary sense of the spread among data given by the range over which the data vary, from the smallest value to the largest value in the data set. The min function identifies the smallest value in a set of numeric elements in a column or row of data. To find the minimum for a set of data, locate your cursor in a spreadsheet cell, activate the Insert Function toolbar button, and select the category: Statistical. In the large window, scroll down the list and select the function: MIN. With your cursor active in the field for Number 1, scroll over the data for which you want to determine the minimum on the spreadsheet beneath the window, and click OK. The minimum value of the numeric items in that range will appear in the spreadsheet cell. Alternatively, the minimum for a set of data can be found directly in Excel. Simply type the function: =min(range of data) in a spreadsheet cell. As with the max function, the minimum for two or more columns or rows of data can also be found.

2.1.7   PERCENTILE

A special class of measures is useful in dividing a data set into proportionate segments. They are quantiles, and we have already introduced one of them, the median.

- The **median** is a quantile that divides a data set into two equally populated halves, with 50% of the data set falling above the median and 50% of the data set falling below the median.
- A **quartile** divides a data set further by splitting the lower half and the upper half in two, so that there are four equally populated quarters of the data set, each containing 25% of the data values.
- A **percentile** divides a data set into 100 equally populated segments, each containing 1% of the data values.

If you have ever taken a national examination, you probably received a scaled score for the exam that was equated to a percentile. A reported score equated to the 87th percentile, for example, means that 87% of the people taking the same test earned scores at or below and 13% of the people taking the test earned scores above the reported score, which establishes a measure of the relative position of the reported score within the entire data set.

To find a percentile for a set of data, locate your cursor in a spreadsheet cell, activate the Insert Function toolbar button, and select the category: Statistical. In the large window, scroll down the list and select the function: PERCENTILE. With your cursor active in the field for Array, scroll over the data for which you want to determine a percentile on the spreadsheet beneath the window. In the field below Array, labeled K, identify the percentile you are interested in, remembering that the percentile is given as a decimal value between 0 and 1. So, for example, if you were interested in the 20th percentile, you would enter 0.2 in the field labeled K.

Example 2.6: Finding the 50th Percentile

To find the 50th percentile of a set of data, we will return to the data for Example 2.1 that were entered in cells A1 through A8 on the Excel spreadsheet.

The Question

A sample of eight different vehicles in the corporate car pool is selected and their city mileages per gallon of fuel are measured, producing the values 19.7, 17.6, 21.4, 18.3, 19.5, 18.2, 19.0, and 18.9 mpg. Find the 50th percentile in the set of data.

Answer

$x = 18.95$

Using Excel

Enter the data in individual cells A1 through A8. With your cursor in cell A9, activate the Insert Function button, select the Statistical category, highlight PERCENTILE, and click OK. Activate your cursor in the field for Array and scroll your cursor over cells A1 through A8 containing the data. Then enter the value 0.5 in the field labeled K, to represent 50% = 0.5 as its decimal equivalent. Click OK. Excel will return the 50th percentile of the set, 18.95, in cell A9.

Discussion

The 50th percentile of the set is 18.95. Half of the values fall above 18.95 and half of the values fall below 18.95. This is the same value we found in Example 2.6 was the median of the set. This verifies that the 50th percentile is the median of the set.

———————

2.1.8 PERMUT

A permutation gives us the number of ways a subset of objects can be taken from a larger set in a particular order. Calculating the total number of ways that can be done in different orders is an important function. Among the top ten dogs in a dog show, for example, first, second, and third prizes can be awarded in $10 \times 9 \times 8$ or 720 different ways. That happens because there are ten dogs who can earn first place. Once first place is awarded, the remaining nine dogs are eligible for second place, and then the remaining eight dogs are eligible for the third place. Algebraically, the total number of ways n items can be arranged in different orders is n factorial, written as $n!$, which is equal to $n \times (n-1) \times (n-2) \times ... \times 1$. The permutation of a subset of r objects selected from a set of n objects is:

$$_nP_r = P(n,r) = \frac{n!}{(n-r)!} = n \times (n-1) \times (n-2) \times ... \times (n-r+1)$$

To find the permutation of n elements taken r at a time in Excel, locate your cursor in a spreadsheet cell, activate the Insert Function toolbar button, and select the category: Statistical. In the large window, scroll down the list and select the function: PERMUT. With your cursor active in the field for Number, type n, the total number of elements in the set. With your cursor active in the field for Number chosen, type r, the number of elements to be selected. Click OK. The value of the permutation will appear in the spreadsheet cell. Alternatively, the permutation of n things taken r at a time can be found directly in Excel. Simply type the function: =permut(n, r) in a spreadsheet cell.

2.1.9   QUARTILE

In addition to finding the median, or 50th percentile, and other percentiles as identified above, we may want to find a particular quartile. Since the second quartile is the median and the fourth quartile is the maximum value in the set of data, the most frequently used quartiles are the first and third quartile. The first quartile is the value below which 25% of the data values fall and above which 75% of the values in the set fall. The third quartile is the value below which 75% of the data values fall and above which 25% of the values in the set fall. One of the simplest procedures to find the first and third quartiles by hand is to simply apply the procedure for finding the location of the median to the lower half of the data set, yielding the first quartile, and to the upper half of the data set, yielding the third quartile.

To find the first quartile in Excel, locate your cursor in a spreadsheet cell, activate the Insert Function toolbar button, and select the category: Statistical. In the large window, scroll down the list and select the function: QUARTILE. With your cursor active in the field for Array, scroll over the data for which you want to determine a percentile on the spreadsheet beneath the window. In the field below Array, labeled Quart, identify the quartile you are interested in, and click OK. In this function, you simply type in 1, 2, or 3 to find the first, second, or third quartile, respectively. You should be aware that, while there is broad consensus on the procedure to use in finding the median or second quartile, some numeric differences may exist in the value Excel identifies as either the first or the third quartile in comparison to other numeric procedures used to find the quartile of interest.

2.1.10   STDEV

The standard deviation is an important measure of spread within a set of numeric data. It is used frequently in statistics to determine how common or how unusual a value is in comparison to other values. When the values under consideration represent the entire population of possible data, the standard deviation of that population is denoted by the symbol σ, or *sigma*. When the values in the data set represent a sample taken from a larger population, the standard deviation for the sample is denoted by the symbol *s*.

To find the sample standard deviation for a set of data, locate your cursor in a spreadsheet cell, activate the Insert Function toolbar button, and select the category: Statistical. In the large window, scroll down the list and select the function: STDEV. With your cursor active in the field for Number 1, scroll over the data for which you want to calculate the standard deviation on the spreadsheet beneath the window and click OK. The value for the standard deviation will appear in the spreadsheet cell. Alternatively, the sample standard deviation for a set of data can be found directly in Excel. Simply type the function: =stdev(range of data) in a spreadsheet cell. As with the other statistical functions, the standard deviation for two or more columns or rows of data can also be found.

Example 2.7: Finding the Sample Standard Deviation

The standard deviation for a set of data is denoted by the symbol s and represents a measure of the spread that exists among the data in the sample.

The Question

A sample of eight different vehicles in the corporate car pool is selected and their city mileages per gallon of fuel are measured, producing the values 19.7, 17.6, 21.4, 18.3, 19.5, 18.2, 19.0, and 18.9 mpg. Find the standard deviation among the mileages in the sample.

Answer

$s = 1.1683$

Using Excel

Enter the data in individual cells A1 through A8. With your cursor in cell A9, activate the Insert Function button, select the Statistical category, highlight STDEV, and click OK. Activate your cursor in the field for Number 1 and then scroll your cursor over cells A1 through A8 containing the data. Click OK. Excel will return the value of the standard deviation, 1.1683, in cell A9. This is equivalent to the procedure outlined in Chapter 1, Example 1.3.

Discussion

The standard deviation among the mileages for the sample of eight cars is $s = 1.1683$ mpg. The standard deviation can be used to talk about what percent of the data values in the set fall, for example, 2 or 3 units of standard deviation on either side of the mean.

2.1.11   STDEVP

In the study and use of statistics, it is important to know whether the value is formed using all elements in the population or whether it is based on a random sample of elements taken from the population. When the data set represents the entire population of all values, the standard deviation for the population is σ, or *sigma*. There are different equations used to calculate a sample standard deviation and a population standard deviation, so Excel makes it clear among its statistical functions which standard deviation is formed based on a sample (=STDEV) and which standard deviation is formed over the entire population (=STDEVP).

To find the population standard deviation for a set of data, locate your cursor in a spreadsheet cell, activate the Insert Function toolbar button, and select the category: Statistical. In the large window, scroll down the list and select the function: STDEVP. With your cursor active in the field for Number 1, scroll over the data for which you want to calculate the standard deviation on the spreadsheet beneath the window and click OK. The value for the standard deviation will appear in the spreadsheet cell. Alternatively, the population standard deviation for a set of data can be found directly in Excel. Simply type the function: =stdevp(range of data) in a spreadsheet cell. As with the other statistical functions, the standard deviation for two or more columns or rows of data can also be found.

2.1.12   VAR

The variance is also an important measure of spread within a set of numeric data. It is the square of standard deviation and, like standard deviation, is used frequently in statistics to determine how common or how unusual a value is in comparison to other values. When the other values in the data set represent a sample taken from a larger population, the variance for the sample is denoted by the symbol s^2.

To find the sample variance for a set of data, locate your cursor in a spreadsheet cell, activate the Insert Function toolbar button, and select the category: Statistical. In the large window, scroll down the list and select the function: VAR. With your cursor active in the field for Number 1, scroll over the data for which you want to calculate the variance on the spreadsheet beneath the window and click OK. The value for the sample variance will appear in the spreadsheet cell. Alternatively, the sample variance for a set of data can be found directly in Excel. Simply type the function: =var(range of data) in a spreadsheet cell. As with the other statistical functions, the variance for two or more columns or rows of data can also be found.

2.1.13   VARP

When the other values in the data set represent all values in a given population, the variance for the population is denoted by the symbol σ^2.

To find the population variance for a set of data, locate your cursor in a spreadsheet cell, activate the Insert Function toolbar button, and select the category: Statistical. In the large window, scroll down the list and select the function: VARP. With your cursor active in the field for Number 1, scroll over the data for which you want to calculate the variance on the spreadsheet beneath the window and click OK. The value for the population variance will appear in the spreadsheet cell. Alternatively, the population variance for a set of data can be found directly in Excel. Simply type the function: =varp(range of data) in a spreadsheet cell. As with the other statistical functions, the population variance for data contained in two or more columns or rows of data can also be found.

2.2   Inserting Selected Math & Trig Functions

The Math & Trig category under the Insert Function toolbar button has nearly 70 different functions listed. We introduce ten of those most useful to routine statistical and financial calculations.

2.2.1   ABS

The absolute value of an integer is the value of the integer without regard to its sign. So the absolute value of +20 and the absolute value of −20 are both 20. The notation for the absolute value of n is $|n|$.

To find the absolute value of an integer in Excel, locate your cursor in a spreadsheet cell, activate the Insert Function toolbar button, and select the category: Math & Trig. In the large window, select the function: ABS. With your cursor active in the field for Number, highlight the spreadsheet cell for which you want the absolute value on the spreadsheet beneath the window, and click OK. The absolute value will appear in the spreadsheet cell. Alternatively, the absolute value of a number can be found directly in Excel. Type either =ABS(cell with the value) or =ABS(the value itself) in a spreadsheet cell.

2.2.2   COMBIN

A combination gives the number of ways a subset of objects can be taken from a larger set regardless of any particular order. A combination of a subset of r objects selected from a set of n objects is smaller than a permutation of the same objects because the order in which the r objects occur is not important in a combination. Since r objects can be arranged in $r!$ different ways, the value of a combination is that of a permutation divided by $r!$.

$$_nC_r \text{ or } C(n,r) = \frac{_nP_r}{r!} = \frac{n!}{(n-r)!r!}.$$

Even though a combination is closely related to a permutation, the combination function is located under the Math & Trig category while the permutation function is located under the Statistical category of the Insert Function toolbar button. So, to find the combination of n elements taken r at a time in Excel, locate your cursor in a spreadsheet cell, activate the Insert Function toolbar button, and select the category: Math & Trig. In the large window, scroll down the list and select the function: COMBIN. With your cursor active in the field for Number, type n, the total number of elements in the set. With your cursor active in the field for Number chosen, type r, the number of elements to be selected. Click OK.

The value of the combination will appear in the spreadsheet cell. Alternatively, the combination of *n* things taken *r* at a time can be found directly in Excel. Simply type the function: =combin(n,r) in a spreadsheet cell.

Example 2.8: Evaluating a Combination

A combination identifies the number of ways a subset of items with a specific characteristic can be selected from a broader population of objects when the order in which they occur is not important. Being able to evaluate a combination can help us establish how likely a particular event is to occur.

The Question

In how many ways can we select three items to test from a total shipment of 26 items?

Answer

$$C (26, 3) = 2600$$

Using Excel

Locate your cursor in a cell on the spreadsheet. Activate the Insert Function button, select the Math & Trig category, highlight COMBIN, and click OK. Activate your cursor in the field for Number and enter the total number of items, 26, from which we can pick. Activate your cursor in the field Number_chosen and enter the number of items you are selecting, 3. Click OK. Excel will return the value of the combination, 2600.

Discussion

There are 2600 unique sets of three items that can be selected from a total number of 26 items. So if we labeled each item with a unique letter A to Z, there would be 2600 different groups of three letters: ABC, ABD, ABE, ... Different orderings of the items are not counted separately, so as long as items A, B, and C are the three items selected to test, that is only counted as one group. So ABC, ACB, BAC, BCA, CAB, and CBA are all considered one group and counted only once in the 2600 unique sets.

2.2.3 INT

Using the nearest integer is important in some statistical computations. The integer function in Excel rounds a number down to the integer at or below the number regardless of the size of the decimal component of its value. So, the integer function returns 5 for the number 5.013 as well as for the number 5.987. The integer function returns −2 for the number −1.7 as well as for the number −1.02.

To apply the integer function in Excel, locate your cursor in a spreadsheet cell, activate the Insert Function toolbar button, and select the category: Math & Trig. In the large window, scroll down the list and select the function: INT. With your cursor active in the field for Number, highlight the spreadsheet cell for which you want the integer on the spreadsheet beneath the window, and click OK. The value of the integer will appear in the spreadsheet cell. Alternatively, the integer can be found directly in Excel. Type either =int(cell with the value) or =int(the value itself) in a spreadsheet cell.

2.2.4 LN

Besides providing a model to describe some logarithmic growth patterns, logarithms play an important role in financial calculations. In exponential notation, a value is a base raised to an exponent, or value = base^{exponent}. The same relationship can be expressed in logarithmic notation as $\log_{\text{base}}$ value = exponent. Stated simply, logarithms are exponents. A frequently used base for logarithms is the number e, which is approximated by 2.71828. A logarithm with a base e is called a natural logarithm, denoted by the symbol ln rather than log. So, $\log_e x = \ln x$, and represents the exponent to which e is raised to equal the value x.

To find the natural logarithm of a value in Excel, locate your cursor in a spreadsheet cell, activate the Insert Function toolbar button, and select the category: Math & Trig. In the large window, scroll down the list and select the function: LN. With your cursor active in the field for Number, highlight the spreadsheet cell containing the value you want for which the natural logarithm on the spreadsheet beneath the window, and click OK. The value of the natural logarithm will appear in the spreadsheet cell. Alternatively, the natural logarithm can be found directly in Excel. Type either =ln(cell with the value) or =ln(the value itself) in a spreadsheet cell.

2.2.5   LOG10

Because our number system is organized on a base of 10, common logarithms are computed on a base of 10. In mathematics, we refer to logarithms in base 10 so frequently that we often do not specify that 10 is the base. So, for example, $\log_{10} 1000 = 3$ is often written simply $\log 1000 = 3$. In Excel, however, we must specify the base of 10 in the function name.

To find the common logarithm of a value in Excel, locate your cursor in a spreadsheet cell, activate the Insert Function toolbar button, and select the category: Math & Trig. In the large window, scroll down the list and select the function: LOG10. With your cursor active in the field for Number, highlight the spreadsheet cell containing the value you want for which the common logarithm on the spreadsheet beneath the window, and click OK. The value of the common logarithm will appear in the spreadsheet cell. Alternatively, the common logarithm can be found directly in Excel. Type either =log10(cell with the value) or =log10(the value itself) in a spreadsheet cell.

Example 2.9: Evaluating a Logarithm, Base 10

Logarithms give us a way to evaluate exponents directly in lieu of relying on trial and error. Logarithms are useful evaluating growth rates in biology and in some areas of finance.

The Question

How many years does it require for $1.00 to become $1.50 at a compound interest rate of 3% compounded annually?

Answer

$$(1.03)^x = 1.50$$
$$x \cdot \log_{10}(1.03) = \log_{10}(1.50)$$
$$x = \frac{\log_{10}(1.50)}{\log_{10}(1.03)} = \frac{0.176091}{0.012837} = 13.71724$$

Using Excel

Enter the value 1.50 in cell A10 on a spreadsheet. Enter the value 1.03 in cell A11. Locate your cursor in cell B10. Activate the Insert Function button, select the Math & Trig category, highlight LOG10, and click OK. Activate your cursor in the field for Number and click on cell A10 to evaluate the log base 10 of 1.50. Click OK. Excel will return the log base 10 of 1.50 as 0.176091 in cell B10. Move your cursor over the lower right corner of cell B10 until your cursor turns from a hollow plus to a solid plus sign. Hold your left mouse button down and drag the function into cell B11. Release the left button. Excel will return the value of the log base 10 of 1.03 as 0.012837 in cell B11. To complete the final step, activate your cursor in cell B12 and type: = B10/B11. When you hit enter, Excel will return the final answer, 13.71724, in cell B12.

Discussion

It will require better than 13.7 years for $1.00 to become $1.50 if left in an account that compounds 3% annually.

————————————————

2.2.6   ROUND

Often what we need is not the exact value of a calculation but a rounded value of the calculation. For example, applying a periodic interest rate to determine an updated account balance requires the calculation be rounded to two digits beyond the decimal point. When an investor submits an amount to purchase shares in a stock or bond fund, the number of shares purchased at the market rate are reported at the thousandths level. The round function in Excel requires two inputs: (1) the number to be rounded and (2) the number of digits beyond the decimal point the function should return. Once the number of digits is specified, Excel evaluates the digit to the right of the number of digits specified and rounds up if that digit is 5 or more and rounds off if that digit is 4 or less.

To apply the rounding function in Excel, locate your cursor in a spreadsheet cell, activate the Insert Function toolbar button, and select the category: Math & Trig. In the large window, scroll down the list and select the function: ROUND. With your cursor active in the field for Number, highlight the spreadsheet cell for which you want the rounded value on the spreadsheet beneath the window. With your cursor active in the field for num_digits, type the number of significant digits you want the number rounded to. Click OK. The rounded value will appear in the spreadsheet cell. Alternatively, the value can be rounded directly in Excel. Type either =round(cell with the value,number of digits) or =round(the value itself,number of digits) in a spreadsheet cell.

2.2.7 SQRT

We often need to take a square root of a value. In statistics, for example, the square root of variance is standard deviation.

To find the square root of a value in Excel, locate your cursor in a spreadsheet cell, activate the Insert Function toolbar button, and select the category: Math & Trig. In the large window, scroll down the list and select the function: SQRT. With your cursor active in the field for Number, highlight the spreadsheet cell containing the value you want for the square root on the spreadsheet beneath the window, and click OK. The value of the square root will appear in the spreadsheet cell. Alternatively, the square root can be found directly in Excel. Type either =sqrt(cell with the value) or =sqrt(the value itself) in a spreadsheet cell.

Example 2.10: Evaluating a Square Root

The square root of a number is the value, when multiplied by itself, equals the given number. So, for instance, the square root of 16 is 4, because $4 \cdot 4 = 16$. The square root is an integer only when the number itself is a perfect square.

The Question

Evaluate the square root of 70.

Answer

$$\sqrt{70} = 8.3666$$

Using Excel

Locate your cursor in a cell on the spreadsheet. Activate the Insert Function button, select the Math & Trig category, highlight SQRT, and click OK. Activate your cursor in the field for Number and enter the number 70. Click OK. Excel will return the value of the square root of 70 as 8.3666.

Discussion

The square root of 70 is approximately 8.3666. That is, if you multiply 8.3666 times itself, you get nearly 70. Some error is introduced due to rounding. The value 8 times itself is 64, and the value 9 times itself is 81. So it makes sense that the square root of 70 should fall between 8 and 9, since 70 falls between 64 and 81.

2.2.8   SUM

Summing columns or rows of data is important in many applications. From simple activities like reconciling an account balance to complex applications like calculating a monthly payment on a loan, the summation function plays a key role in basic calculations. In statistics, the sum is the basis for calculating both the mean and the variance.

In acknowledging the frequency with which users of Excel need to access the sum for a stream of data, Excel programmers placed a separate AutoSum button on the upper toolbar: $\boxed{\Sigma}$. To find the sum of a set of data, locate your cursor in a spreadsheet cell, activate the AutoSum button, then scroll over the data for which you want to calculate the sum on the spreadsheet beneath the window, and click OK. The value for the sum will appear in the spreadsheet cell. Alternatively, you can access the summation function within the Insert Function toolbar button under the Math & Trig category. Scroll down and select the SUM function. With your cursor active in the field for Number 1, scroll over the data for which you want to calculate the sum on the spreadsheet beneath the window and click OK. The value for the sum will appear in the spreadsheet cell.

The sum for a set of data can also be found directly in Excel. Simply type the function: =sum(range of data) in a spreadsheet cell. As with the other Excel functions, the sum for data contained in two or more columns or rows of data can also be found.

Example 2.11: Finding the Sum

To find the sum of a set of data, we will return to the data for Example 2.1 that were entered in cells A1 through A8 on the Excel spreadsheet.

The Question

A sample of eight different vehicles in the corporate car pool is selected and their city mileages per gallon of fuel are measured, producing the values 19.7, 17.6, 21.4, 18.3, 19.5, 18.2, 19.0, and 18.9 mpg. Find the sum of the sampled values.

Answer

$$\sum x = 152.6$$

Using Excel

Enter the data in individual cells A1 through A8. With your cursor in cell A9, activate the Insert Function button, select the Math & Trig category, highlight SUM, and click OK. Activate your cursor in the field for Number1 and scroll your cursor over cells A1 through A8 containing the data. Click OK. Excel will return the sum of the set, 152.6, in cell A9.

Discussion

The total of the eight values added together is 152.6. As noted in Chapter 1, Example 1.1, the sum of a set of numbers can be the first step in finding the average for the numbers.

2.2.9   TRUNC

Occasionally what we need is neither a rounded nor the nearest integral value of a calculation but the last value of the calculation specified to a certain number of decimal places. The truncating function in Excel levels a number back to the last value given a certain number of decimal places. So, the truncating function for two decimal values returns 5 for the number 5.013 as well as for the number 5.987. The truncating function for two decimal values returns −2 for the number −1.99 and −3 for the number −2.05. The truncating function for one decimal value returns −1.9 for the number −1.99 but −1.4 for the number −1.46. The truncating function in Excel requires two inputs: (1) the number to be rounded and (2) the number of digits beyond the decimal point the function should evaluate.

To apply the truncating function in Excel, locate your cursor in a spreadsheet cell, activate the Insert Function toolbar button, and select the category: Math & Trig. In the large window, scroll down the list and select the function: TRUNC. With your cursor active in the field for Number, highlight the spreadsheet cell for which you want the rounded value on the spreadsheet beneath the window. With your cursor active in the field for num_digits, type the number of significant digits you want to be considered. Click OK. The truncated value will appear in the spreadsheet cell. Alternatively, the value can be rounded directly in Excel. Type either =trunc(cell with the value,number of digits) or =trunc(the value itself,number of digits) in a spreadsheet cell.

2.3   Inserting Selected Financial Functions

The Financial category under the Insert Function toolbar button has nearly 90 different functions listed. Many of the automated functions are designed to conduct calculations for complex financial applications. We introduce a few of the more frequently used financial functions.

2.3.1   EFFECT

When interest is assessed on the balance of an account on any other than an annual basis, the annual interest rate on the account does not allow for the compounding effect taking place. The effective interest rate, r_E, is the simple interest rate that, when applied to the balance of the account, would generate the same amount of interest as generated by the annual interest rate compounded over the year. Because of compounding, the effective interest rate is higher than the stated annual rate. Credit card accounts usually announce both the annual and the effective rates of interest.

To find the effective interest rate in Excel, locate your cursor in a spreadsheet cell, activate the Insert Function toolbar button, and select the category: Financial. In the large window, scroll down the list and select the function: EFFECT. With your cursor active in the field for Nominal_rate, enter the stated annual interest rate as a decimal equivalent. With your cursor active in the field for Npery, enter the number of compounding periods per year and click OK. The value of the effective interest rate will appear in the spreadsheet cell. Alternatively, the effective interest rate can be found directly in Excel. Type =effect(stated annual interest rate,number of compounding periods per year) in a spreadsheet cell.

Example 2.12: Finding the Effective Interest Rate

To find the simple interest rate that, when applied to the balance of an account, would generate the same amount of interest as generated by the annual interest rate compounded over the year, we will use the effective interest rate function in Excel.

The Question

A store account identifies 18% interest rate compounded monthly. Find the effective interest rate that is equivalent to this compounded annual rate.

Answer

$r_E = 0.195618$ or approximately 19.56%

Using Excel

Locate your cursor in a cell on the spreadsheet. Activate the Insert Function button, select the Financial category, highlight EFFECT, and click OK. Activate your cursor in the field for Nominal rate and enter the number 0.18, the stated annual simple interest rate. Activate your cursor in the field for Npery and enter the number 12, the number of interest or payment periods in a year. Click OK. Excel will return the value of the effective rate of return, 0.195618, in the cell.

Discussion

If you divide the annual interest rate 0.18 by 12, the monthly interest rate is 0.015. When you raise (1 + 0.015) to the power of 12, you will get: $(1 + 0.015)^2 = 0.195618$, which is the effective interest rate for that account.

─────────────────────

2.3.2   FV

The future value of a stream of constant payments made each interest period is called the future value of an annuity. It represents the value of the account at the end of the final deposit and is comprised of the sum of the periodic payments plus all compound interest accrued over the life of the account.

To find the future value of a stream of payments in Excel, locate your cursor in a spreadsheet cell, activate the Insert Function toolbar button, and select the category: Financial. In the large window, scroll down the list and select the function: FV. With your cursor active in the field for Rate, enter the periodic interest rate as a decimal equivalent. If, for example, the interest rate is 6% compounded monthly, the periodic interest rate would be $0.06 \div 12 = 0.005$. With your cursor active in the field for Nper, enter the total number of compounding periods over the life of the account. For example, you would enter 36 for an account of 3 years duration with monthly compounding, since $3 \times 12 = 36$. With your cursor active in the field for PMT, enter as a negative value the constant amount deposited with each payment. Not required are entries in the fields for Pv or Type. Click OK. The future value of the stream of payments will appear in the spreadsheet cell. Alternatively, the future value can be found directly in Excel. Type =fv(periodic interest rate,number of interest periods over the life of the account,value of the constant payment made to the account as a negative value) in a spreadsheet cell.

Example 2.13: Finding the Future Value of a Stream of Equal Payments

To find the future value of an ordinary annuity, we will use Excel's built-in function, FV.

The Question

Tran is going on active duty for a 24-month assignment overseas. He anticipates he will be able to save $8.99 a month by not streaming first-run movies while he is on active duty. Instead of incurring the standard charge to his credit card during the 24-month period, Tran elects to authorize his bank to withdraw the amount from his checking account as an automatic payment into a savings account that earns 3% annual interest rate compounded monthly. How much will be in the account at the end of the 24-month duty period when he returns home?

Answer

$FV = \$222.08$

Using Excel

Locate your cursor in a cell on the spreadsheet. Activate the Insert Function button, select the Financial category, highlight FV, and click OK. Activate your cursor in the field for Rate and enter the number 0.03/12, the annual interest rate converted to a periodic interest rate. Activate your cursor in the field for Nper and enter the number 24, the number of interest or payment periods over which the payment will be made. Activate your cursor in the field for Pmt and enter the value −8.99, the amount of the monthly payment. Click OK. Excel will return the future value of the account, $222.08, in the cell.

Discussion

The future value of the stream of 24 payments each for $8.99 into Tran's account earning 3% interest will be worth $222.08 at the end of

24 months. The payments themselves account for $8.99 × 24 = $215.76, and there was a total of $6.32 earned in interest over the 24-month period.

2.3.3   NPV

The net present value (NPV) function produces the present value of a stream of irregular payments assuming a given discount rate. The argument for the function is in the form of =NPV(discount rate, range of future payments).

2.3.4   PMT

The periodic payment needed to pay off the present value of a loan over a specific number of payments and interest rate is a useful calculation for consumers. It includes payment to the principal and interest due on the outstanding balance, and is calculated so that the constant payment over the specified number of payments fully covers the outstanding balance and its interest.

To find the periodic payment in Excel, locate your cursor in a spreadsheet cell, activate the Insert Function toolbar button, and select the category: Financial. In the large window, scroll down the list and select the function: PMT. With your cursor active in the field for Rate, enter the periodic interest rate as a decimal equivalent. If, for example, the interest rate is 6% compounded monthly, the periodic interest rate would be $0.06 ÷ 12 = 0.005$. Alternatively, you can enter the annual interest rate, the division sign "/" followed by 12. With your cursor active in the field for Nper, enter the total number of compounding periods over the life of the account. For example, you would enter 36 for an account of 3 years duration with monthly compounding, since $3 × 12 = 36$. With your cursor active in the field for PV, enter the total amount due. Usually that is the outstanding balance of the asset being purchased with payments made over time. Click OK. The value of the payment will appear in the spreadsheet cell. Alternatively, the effective interest rate can be found directly in Excel. Type =pmt(periodic interest rate,total number of periods over the life of the loan,outstanding balance of the asset) in a spreadsheet cell.

Example 2.14: Finding the Periodic Payment

We will use the programmed function PMT to find the amount of a payment required each period to pay off in a stated number of payments the present value owed, given an assumed interest rate.

The Question

Mario and Luz buy a living room furniture set which, with tax and delivery, totals $3895.92. They put $300 down and plan to pay the remaining balance over 36 monthly payments, with 2½% annual interest rate. How much each month will they need to plan to pay it off in the stated amount of time?

Answer

$PMT = \$103.78$

Using Excel

Locate your cursor in a cell on the spreadsheet. Activate the Insert Function button, select the Financial category, highlight PMT, and click OK. Activate your cursor in the field for Rate and enter the number 0.002083 or 0.025/12, the annual interest rate converted to a periodic interest rate. Activate your cursor in the field for Nper and enter the number 36, the number of interest or payment periods over which the payment will be made. Activate your cursor in the field for PV and enter the value 3595.92, the amount they currently owe when they take possession of the furniture set, the amount of the monthly payment. Click OK. Excel will return the value of the monthly payment required to pay the debt off in 36 monthly payments with 2½% annual interest rate, $103.78, in the cell.

Discussion

The total value of the stream of payments will equal $103.78 × 36 = $3,736.08. By making payments, Mario and Luz will have paid $140.16 in total interest above the outstanding amount of $3,595.92 due when they took possession of the asset.

2.3.5 PV

The present value of a stream of constant, equally-timed payments made into the future at a set interest rate each interest period is called the present value of an annuity. It represents the lump sum that would need to be deposited today to equal a stream of constant payments into the future earning the same interest rate.

To find the present value of a stream of payments in Excel, locate your cursor in a spreadsheet cell, activate the Insert Function toolbar button, and select the category: Financial. In the large window, scroll down the list and select the function: PV. With your cursor active in the field for Rate, enter the periodic interest rate as a decimal equivalent. If, for example, the interest rate is 6% compounded monthly, the periodic interest rate would be $0.06 \div 12 = 0.005$. With your cursor active in the field for Nper, enter the total number of compounding periods over the life of the account. For example, you would enter 36 for an account of 3 years duration with monthly compounding, since $3 \times 12 = 36$. With your cursor active in the field for Pmt, enter as a negative value the constant amount deposited with each payment. Not required are entries in the fields for Fv or Type. Click OK. The present value of the stream of payments will appear in the spreadsheet cell. Alternatively, the present value can be found directly in Excel. Type: =pv(periodic interest rate,number of interest periods over the life of the account,value of the constant payment made to the account as a negative value) in a spreadsheet cell.

Example 2.15: Finding the Present Value of a Stream of Equal Payments

To find the present value of an ordinary annuity, we will use Excel's built-in function, PV.

The Question

Sharnell won a state-sponsored lottery that guarantees her $2,500 per month for 25 years. Assuming a constant annual interest rate of 2% over the 25 years, how much in a lump sum would have to be deposited into an account today earning 2% annual interest compounded monthly to equal this stream of payments?

Answer

$$PV = \$589,825.27$$

Using Excel

Locate your cursor in a cell on the spreadsheet. Activate the Insert Function button, select the Financial category, highlight PV, and click OK. Activate your cursor in the field for Rate and enter the number 0.02/12, the annual interest rate converted to a periodic interest rate. Activate your cursor in the field for Nper and enter the number 300, the number of interest or payment periods over which the payment will be made. Alternatively, you can enter 12*25 in the Nper field, since there will be 12 payments a year made over 25 years. Activate your cursor in the field for Pmt and enter the value −2500, the amount of the monthly payment. Click OK. Excel will return the present value of the stream of future payments, $589,825.27, in the cell.

Discussion

The present value of the stream of 300 monthly payments each for $2,500 to Sharnell assuming 2% interest is $589,825.27. The payments themselves would have totaled $2500 × 300 = $750,000 over time, which is being discounted $160,174.73 to today in recognition of the interest that will be generated over the years that the account funds the stream of payments.

2.3.6   RATE

The periodic interest rate for a loan can be computed based on the periodic payment, the number of payment periods in the life of the loan, and the present value of the asset being purchased with payments made over time.

To find the periodic interest rate in Excel, locate your cursor in a spreadsheet cell, activate the Insert Function toolbar button, and select the category: Financial. In the large window, scroll down the list and select the function: RATE. With your cursor active in the field for Nper, enter the total number of compounding periods over the life of the account. For example, you would enter 36 for an account of 3 years duration with monthly compounding, since $3 \times 12 = 36$. With your cursor active in the field for Pmt, enter as a negative value the constant amount deposited with each payment. With your cursor active in the field for Pv, enter the outstanding balance owed on the asset being purchased with payments made over time. Not required are entries in the fields for Fv or Type. Click OK. The periodic interest rate will appear in the spreadsheet cell. Alternatively, the present value can be found directly in Excel. Type: =rate(periodic interest rate,value of the constant payment made to the account as a negative value,present value of the asset being purchased with payments made over time) in a spreadsheet cell.

CHAPTER 3

Working with Data

As reviewed in the prior chapter, Microsoft Excel has powerful quantitative tools built into its capabilities under the Insert Function toolbar button. This chapter will introduce additional capabilities that greatly enhance the user's ability to summarize and analyze a set of data. While one of the goals of this chapter is to preview the tools in Excel's Data Analysis ToolPak, the reader may want to expand their facility with basic statistical analysis by reviewing a companion publication, *Working with Sample Data: Exploration and Inference*, in the Business Expert Press Quantitative Approaches to Decision Making collection.

3.1   Installing the ToolPak

Excel for Mac systems does not offer the Data Analysis ToolPak. However, if the Apple machine runs Windows emulation software and has Office for the PC installed, it can run the Data Analysis ToolPak.

Excel is often only partially installed on Windows-based machines. To verify installation of the Data Analysis ToolPak on newer versions of Excel, activate the Data tab on the toolbar at the top of your screen. Data Analysis tools should be available on the far right of the Data ribbon. If it does not appear on the ribbon, you will need to activate it. To add the Data Analysis ToolPak in Excel 2010, activate the Files tab and select Options. On the left side of the Excel Options window, activate the item labeled: Add-Ins. The last line of the new window just activated reads: Manage: Excel Add-ins. Click Go and make sure on the next window that the Analysis ToolPak option is selected. Click OK. You should now see Data Analysis appear at the far right of the Data ribbon at the top of your Excel spreadsheet.

For older versions of Excel, click the windows icon in the upper left corner of your screen. Select the Excel Options tab at the bottom of the

window and then select Add-Ins from the list on the left of the new window. Highlight the Analysis ToolPak entry and click the OK button. Your
computer will connect to Microsoft's web page and install the program.
If you are operating Microsoft Excel 97 or XP, you may need to install
the toolkit from the original compact disc (CD) used to load Excel onto
your machine.

3.2   Using the ToolPak and Interpreting Results

In newer versions of Excel, the ToolPak is located on the Data ribbon at
the top of the Excel window. In Excel 97 or Excel XP, Data Analysis is
contained in the pull-down Tools menu. The analysis tools listed are programmed to automate complex, multi-step analyses.

Knowing something about the data in question is important to their
numeric summary. The two main types of data are qualitative and quantitative. Qualitative data arise in settings where sampled elements are
classified by a key attribute and assigned to categories, such as defective
versus acceptable products, commodities whose prices rose as opposed to
remained the same or even fell, or the number of sales made using credit
rather than debit card, check, or cash. Statistical analysis of these qualitative, or categorical, variables is limited largely to counts of how many
elements sampled fall into each category.

Quantitative variables capture data arising from measurements that
include, for example, the dimensions of time, distance, length, weight,
volume, rates, percent, and monetary value. Sample data generated by
measurements take on a value along a number line. Because the data are
in the form of numerical values, there is a richer set of techniques for
analyzing these data, including several of the statistical functions cov-
ered in Chapter 2, like average, median, and standard deviation. A more
complete discussion of discrete and continuous variables can be found in
Section 3.3 of this chapter and in Chapter 1 of the companion publication, *Working with Sample Data: Exploration and Inference*, in the Business
Expert Press Quantitative Approaches to Decision Making collection.
The remainder of this section examines some tools in Excel for analyzing
data collected for quantitative variables.

3.2.1   Descriptive Statistics

One of the most useful tools in the ToolPak is Descriptive Statistics. What is important for the appropriate use of the Descriptive Statistics tool is that the data to be summarized represent quantitative measurements taken on individual sampled elements. The tool cannot be used on data that are already summarized or on qualitative data that are generated by counts of elements sorted into categories. The tool requires quantitative data and, from the data, produces an output range of 13 standard values, including estimates of the center (the mean, the median, the mode) and measures of spread (the range, the minimum, the maximum, the sample standard deviation, the sample variance) as well as the sum of all data, and the number, or count, of values summarized. While each of the values presented in the output can be derived individually with equations and formulas, using the Descriptive Statistics tool can give the user a very quick overview of the set of data in question. The tool is applied to quantitative variables and can be used to summarize data with large sample sizes as quickly as with small sample sizes. We will not address kurtosis or skewness here, even though they are measures routinely reported with Descriptive Statistics. Please see Example 3.1, Using the Descriptive Statistics Tool

Example 3.1: Using the Descriptive Statistics Tool

The Question

A sample of eight different vehicles in the corporate car pool is selected and their city mileages per gallon of fuel are measured, producing the values 19.7, 17.6, 21.4, 18.3, 19.5, 18.2, 19.0, and 18.9 mpg. Find the descriptive statistics for the sample.

Using Excel

Enter the label City MPG in cell A1 and the data in individual cells A2 through A9. Activate the data ribbon at the top of the screen and select Data Analysis at the far right end of the ribbon. Highlight Descriptive Statistics on the list of analysis tools. Click OK. On the Input Range,

Answer

Table 3.1. Descriptive Statistics, City Mileage for Corporate Car Pool

	A	B	C
1	City MPG	*City MPG*	
2	19.7		
3	17.6	Mean	19.075
4	21.4	Standard Error	0.413067791
5	18.3	Median	18.95
6	19.5	Mode	#N/A
7	18.2	Standard Deviation	1.168332145
8	19	Sample Variance	1.365
9	18.9	Kurtosis	1.556499388
10		Skewness	1.019849337
11		Range	3.8
12		Minimum	17.6
13		Maximum	21.4
14		Sum	152.6
15		Count	8

scroll your cursor over cells A1 through A9. Notice that the button for Grouped By Columns is automatically selected. If your data are listed in a row, the button for Grouped By Rows will be selected automatically when you scroll your cursor over the row containing your data. Click the selection Labels in First Row to indicate to Excel that the contents of cell A1 should be used as the label for the output. If there is no label in the lead cell of your data list, leave the selection Labels in First Row unselected. Select Output Range and click your cursor in the field to the right. Either click your cursor on the spreadsheet cell B1, or type the entry B1 into the field. Click the selection Summary Statistics and then click OK. Excel will return the output shown above.

Discussion

The mean, or average, mileage for the sample of eight cars is $\bar{x}$ = 19.075 mpg. The middle value, the median, is 18.95 mpg. There is no value that appears more than once, so there is no mode (#N/A). The sample standard deviation is $s = 1.168332$ and the sample variance is $s^2 = 1.365$. The maximum mileage is 21.4 mpg, the minimum mileage is 17.6 mpg, and their difference is the range of values represented in the sample, which is $21.4 - 17.6 = 3.8$ mpg. The total of all values is the sum of 152.6 and the count of values summarized is $n = 8$. The standard error is useful in comparing sample means. It is the standard deviation among sample means of the same sample size drawn from the same population and is the sample standard deviation, s, divided by the square root of the count, n, $\dfrac{s}{\sqrt{n}} = \dfrac{1.168332}{\sqrt{8}} = 0.413068$. These summary values coincide with sample values found in earlier examples using the same data.

3.2.2 Building Histograms

The graphic summary of quantitative variables can be presented in various ways. One of the most frequently used displays of quantitative variables is the histogram. A histogram is a column chart, where the height of each column on the vertical axis represents the frequency with which data occur in consecutive, nonoverlapping classes and whose columns are contiguous to convey the fact that there is a continuous scale on the horizontal axis.

To build a histogram, we must first choose an approximate number of classes for the data. While there is no "best" number of classes to divide the data among, a reasonable estimate of k, the number of classes to use, can be formed based on n, the number of values in a data set. In particular, begin with a value k such that $2^k > n$. So, for example, if we had a set of 50 data values, $2^5 = 32$ which is less than 50, but 2^6 is 64, which is greater than 50. We would begin by trying 6 classes to divide the data among.

Once you estimate the number of classes to use, divide that number into the range of the data to find the approximate class interval. Round this estimate to a convenient value; most people count and think in multiples of 2, 5, and 10. Determine the lower class limit for the first class by selecting a convenient number that is smaller than the lowest data value in the set. Determine the other class limits by repeatedly adding the class width to the previous class limit. Be sure to mark the midpoint of each class interval on the horizontal axis.

How to Build a Histogram

1. **Number of classes**

   Choose an approximate number of classes, k, for your data.

2. **Estimate the class interval**

   Divide the approximate number of classes *(from Step 1)* into the range of your data to find the approximate class interval, where the range is defined as the largest data value minus the smallest data value.

3. **Determine the class interval**

   Round the estimate *(from Step 2)* to a convenient value.

4. **Lower Class Limit**

   Determine the lower class limit for the first class by selecting a convenient number that is smaller than the lowest data value.

5. **Class Limits**

   Determine the other class limits by repeatedly adding the class width *(from Step 2)* to the previous class limit, starting with the lower class limit *(from Step 3)*.

6. **Define the classes**

   Use the sequence of class limits to define the classes.

   Be sure to label the class midpoint on each class on the horizontal axis.

7. **Develop the frequency of values in each class**

   Plot the frequency, or number of values, that occur in each class as the height of the column on the vertical axis. Excel is quite helpful in accomplishing this step.

Excel's Analysis ToolPak contains a tool that automates the creation of a histogram. Caution should be used in activating it, however, because it can produce class intervals and limits that are not of convenient dimensions. To control the establishment of classes, or bins as they are referred to in Excel, we recommend taking time to define the class intervals and limits prior to activating the histogram tool, then referencing the defined classes as input while using the tool.

Example 3.2: Using the Histogram Tool

The Question

Highway mileage ratings for a sample of 67 model 2011 vehicles are shown in Table 3.2. Use a histogram to graphically summarize the data.

Table 3.2. *Highway Mileage Ratings, Model 2011 Vehicles*[1]

	A	B	C	D	E	F
1			**Highway**	**Mileage**		
2	30	28	28	20	30	21
3	29	28	28	20	30	34
4	29	25	28	34	30	35
5	31	28	28	35	27	30
6	24	25	27	38	26	31
7	22	28	26	37	25	28
8	26	28	24	40	26	28
9	29	28	26	31	22	
10	19	27	24	29	23	
11	18	26	26	25	28	
12	27	26	20	26	28	
13	28	25	20	23	25	

Answer

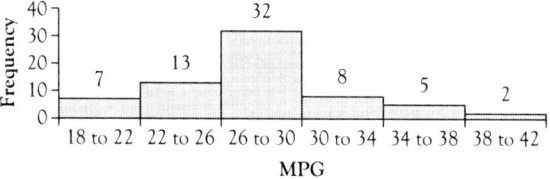

Figure 3.1. *A histogram: Highway mileage for 2011 model US vehicles.*[2]

Using Excel

Enter the data in cells A1 through A67. Run descriptive statistics for the data set using the Analysis ToolPak. Note that the maximum value is 40, the minimum value is 18, and the range is 22. The value of 2^6 is 64 and

2^7 is 128. Consider using 7 classes because 67 is less than 2^7. The range divided into 7 classes yields an estimated interval width of just over 3. Round the estimate to 4. If we start with a lower class limit of 18, we can cover the data set with 6 classes as shown in Figure 3.1. Please note: the suggestion of 7 classes is just that. It is a starting place, not a final answer to the number of classes to use in summarizing the data.

In order for the classes to be adjacent, but not overlapping, we set the first bin to have an upper limit of 21.99. Enter the value 21.99 in cell B1. In cell B2, enter the equation: =B1+4 and copy the equation down through cell B6. You should see the values 21.99, 25.99, 29.99, 33.99, 37.99, and 41.99. These will serve as the upper limits of the six classes the data will be collapsed into. Since you want the value 22 to count in the second class, 22 to 26, we list the upper limit of the first class to be just below 22.

Activate the data ribbon at the top of the screen and select Data Analysis at the far right end of the ribbon. Highlight Histogram on the list of analysis tools. Click OK. On the Input Range, scroll your cursor over cells A1 through A67. In the Bin Range, scroll your cursor over cells B1 through B6. Do *not* mark labels, since we will address that later. Click Output Range, and activate your cursor in the field. Click on cell C1. At the very bottom of the window, highlight Chart Output. Click OK. Excel will return a frequency distribution of the data summarized in cells C1 through D8 in the classes you dictated in the bin range. In cells F1 through K10 will appear a graphic histogram.

Undoubtedly you will want to improve the basic graph.

1. Consider removing the legend. By removing the legend, you will expand the area of the graphic allocated to the depiction of the data.

2. Remove the More category. To delete the More category, click once on the graph. Note that the input data in cells C1 through D8 are highlighted. Grab the lower right-hand edge of the border and drag it to cell D7. The More category should no longer show on the histogram.

3. Reduce the gap width between the columns. To do so, double click your cursor on a column in the histogram. Run the gap width to zero. Should you want to change the color of the

columns, that can be done by selecting Fill in the left margin. Select Solid Fill on the right, and Colors to open a color chart to individualize the color for the columns. Before you exit that window, you may want to consider putting a black border around each column to distinguish the boundaries of each. If so, select Border Color in the left margin. Highlight black and click OK.

4. Formalize the bin labels. Instead of 21.99, in cell B1, type: 18 to 22. In cell B2, type: 22 to 26. In cell B3, type: 26 to 30. In cell B4, type: 30 to 34. In cell B5, type: 34 to 38. In cell B6, type: 38 to 42. Those labels should now appear on the bottom axis of the histogram. If the labels do not appear all on one line of text in the graph, highlight the graph, grab the middle of the right edge of the graph and extend the histogram to cover from cell F1 to cell F10. Should you want to increase the font size of the labels, make them bold, or do both, you may need to make the histogram even bigger.

5. Formalize the label for the horizontal axis. To replace the word "Bin," click the word "Bin" to highlight it and type: MPG. Hit Enter. The axis label MPG should now appear along the bottom axis below the class labels. You may want to increase the font size of the axis label or make it bold. That can be done while the label is highlighted. Click OK.

6. Formalize the label for the histogram. To replace the word "Histogram," click the word Histogram to highlight it and type: Highway Mileage. Hit Enter. The title of the histogram should now read Highway Mileage. Again, you may want to increase the font size of the chart title or make it bold. That can be done while the title is highlighted. Click OK.

7. Consider adding data labels above each column. To do so, right click on a column so that all columns are highlighted. Select Add Data Labels. Click OK.

Discussion

The modal class, or most frequently occurring class, of highway mileage is 26 to 30 mpg, which, at a class frequency of 32, clearly dominates the frequency of any other class. The highway mileages are slightly skewed to the right. That is, the distance from the midpoint of the modal class, 28, to the upper limit of the highest class (28 to 42, or 14) is greater than the distance from the midpoint of the modal class to the lower limit of the smallest class (28 to 18, or 10).

3.2.3   Finding Rank and Percentile

Using the Rank and Percentile tool in the Analysis ToolPak generates a four-column output with columnar titles: Point, Column 1, Rank, Percent. The first column of the output is entitled Point and replicates the original data list. The second column of the output is entitled Column 1 and contains a list of all data sorted from the maximum to the minimum value for the data set. The third column of the output is entitled Rank and contains the position of each of the data elements from 1 to n, where n is the size of the sample. Of particular interest here is that, for repeated values, the rank represents the median value of the rank for the point's place in the overall list. So, for example, the mileage value of 30 mpg appeared five times in the original list of data and occupied the 9th, 10th, 11th, 12th, and 13th positions when the original data were arranged in decreasing order. The median rank among those five ranks is $(9 + 13)/2 = 11$, which is the third rank of the five, or 11th. Excel assigned the tied rank of 11th to each of the five mileages. For second example, the next mileage of 29 mpg appeared four times in the original list of data and occupied the 14th, 15th, 16th, and 17th positions when the original data were arranged in decreasing order. Excel assigned the tied rank of $(14 + 17)/2 = 15.5$, which was then rounded up to the 16th position. Excel assigned the tied rank of 16th to each of the four mileages. The fourth and final column of the output is entitled Percent and reflects the percentile, the percent of all data elements that fall at or below that value when arranged in decreasing order.

3.2.4   Other Analysis Tools

We have just reviewed three of the analysis tools in Excel's Data Analysis ToolPak. While there are a total of 19 tools available, many of them are complex tools beyond the scope of this book. Some of them relate to basic statistical analysis and are used and discussed in the companion publication, *Working with Sample Data: Exploration and Inference.* ANOVA: Single Factor, ANOVA: Two-Factor With Replication, and ANOVA: Two-Factor Without Replication all relate to inferential tests of means taken from multiple populations and are discussed in Chapter 6 of that publication. The *F*-Test Two-Sample for Variances involves inferential tests for the equality of variances from two populations, most useful in determining which *t*-test to conduct in determining differences in their means, and is discussed in Chapter 5 of that publication. The regression tool determines whether two or more variables are related and is discussed in its simplest form in Chapter 8 of that publication. The three *t*-tests automate the comparative analysis of two population means and are discussed in Chapter 5 of that publication, as is the *z*-Test Two Samples for Means.

3.3   Generating Graphics

Graphic summaries provide a picture of key data that can be used to elicit important questions, fuel understanding, and facilitate communication. Important data tell stories worth listening to and worthy of retelling. Good graphics present the actual data and show causality, multiple comparisons, multiple perspectives, the effects of the processes that lead to their creation, or the effects of subsequent changes made to those processes. They should reinforce the reason the data are significant.

3.3.1   Working with Qualitative Data and Grouped Quantitative Data

As discussed at the beginning of Section 3.2, the analysis of data that assess qualitative variables is limited to assigning the observations to categories and counting the occurrences in each category. Visual summaries of data for qualitative variables by category can be achieved in a column chart or some variant of a column chart, including side-by-side column charts or stacked column charts. Recall that in the construction of histograms, we grouped quantitative data values into classes, which are a type of category. Hence, the techniques for displaying category counts for qualitative variables also apply to the data that have been assigned to classes or categories.

Example 3.3: Forming a Column Chart with a Qualitative Variable

The Question

Use the data in Table 3.3 to develop and interpret a side-by-side column chart showing the retail sales over time of domestic and imported new passenger cars.

Table 3.3. Retail Sales of New Passenger Cars in the United States (in Thousands of Units), 1970–2010[3]

	A	B	C	D	E	F	G	H	I	J
1		1970	1975	1980	1985	1990	1995	2000	2005	2010
2	Domestic	7,119	7,053	6,580	8,205	6,919	7,114	6,762	5,473	3,792
3	Imports	1,280	1,572	2,369	2,775	2,384	1,506	2,016	2,187	1,844

Answer

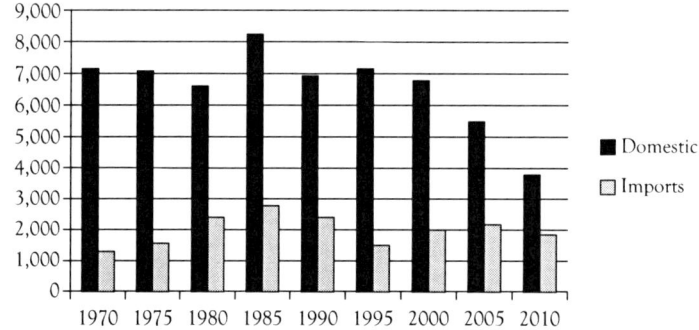

Figure 3.2. A side-by-side column chart: Retail new passenger car sales in the United States (in thousands of units), 1970–2010.[4]

Using Excel

Enter the data as shown in Table 3.3, leaving cell A1 blank. In cell A2, enter the category label: Domestic. In cell A3, enter the category label: Imports. In cell B1, enter the year label 1970, and continue, increasing the year label by 5 until you have entered 2010 in cell J1. Enter the appropriate numbers for the selected category and year as shown in Table 3.3.

Highlight the block of cells from A1 through cell J3 entered from Table 3.3. Select the Insert tab at the top of the Excel window. Select Column and click 2-D Column, left-most icon for a side-by-side column chart.

Again you will probably want to improve the basic graph. To insert a chart title, select Layout and highlight Chart Title. Type in the chosen title for the chart.

Discussion

In the 1970s, there was a growth in the purchase of foreign versus domestic cars due to the increase in the price of gasoline. In 1985, sales of cars increased due to the rising value of the dollar. Between 1985 and 1995, the share of domestic vehicles sold increased due in large part to the increased popularity of SUVs. However, in the last decade, the purchase of US cars began to drop due to the increase in gasoline prices, when popularity of comparatively smaller import cars regained market share. Sales of both domestic and import cars dropped noticeably in 2010 due to the global recession.

The side-by-side column chart above allows the reader to examine the patterns in both categories, domestic and imports, displayed across individual years. When the patterns across individual years are less important than the combined pattern exhibited over the entire period of time, a stacked column chart is more appropriate.

Example 3.4: Forming a Stacked Column Chart with Qualitative Variables

The Question

Use the data in Table 3.3 to develop and interpret a stacked column chart showing the sales over time of domestic and imported new passenger cars.

Answer

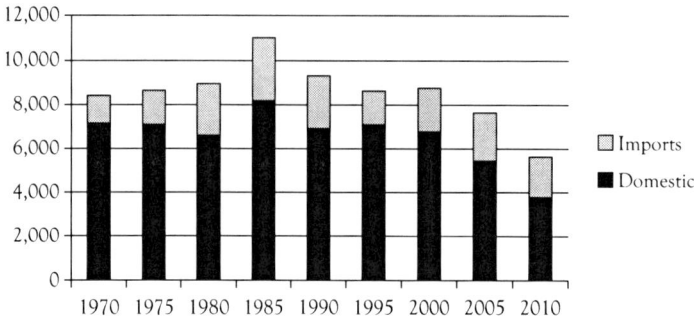

Figure 3.3. A stacked column chart: Retail new passenger car sales in the United States (in thousands of units), 1970–2010.[5]

Using Excel

Highlight the block of cells from A1 through cell J3 entered from Table 3.3. Select the Insert tab at the top of the Excel window. Select Column and click 2-D Column, left-most icon for a side-by-side column chart.

Again you will probably want to improve the basic graph. To insert a chart title, select Layout and highlight Chart Title. Type in the chosen title for the chart.

Discussion

While it is just as easy to compare domestic retail sales over time in Figure 3.3 as in Figure 3.2, Figure 3.3 provides less direct evidence of comparative changes in retail sales of imported passenger cars. Although the evidence is there, visual comparisons in the retail sales of imports over time are more difficult than in the side-by-side chart provided in Figure 3.2. What is more readily available in Figure 3.3 is the behavior of retail sales of all new passenger cars over time, regardless of origin. Retail sales of new passenger cars was remarkably stable over time, with the exceptional high in 1985 resulting from the increased value of the dollar and the exceptional low in 2010 resulting from the global recession.

When the messages to convey with the data are comparative changes in categories over time, a pie chart may be an effective graphic tool to use.

Example 3.5: Forming Pie Charts

The Question

As various agencies project service demands into the future, demographic changes are often central to those projections. Use the data in Table 3.4 to develop and interpret two pie charts for age groups in the US population in 2000 and 2010.

Table 3.4. *US Population by Selected Age Groups (US Census Bureau)*[6]

	A	B	C
1	US Population, Age Groups: 2000 and 2010		
2		2000	2010
3	Under 18 yrs	25.7%	24.0%
4	18−44 yrs	39.9%	36.5%
5	45−64 yrs	22.0%	24.0%
6	65+ yrs	12.4%	13.0%

Using Excel

Enter the data as shown in Table 3.4, leaving cell A1 blank. Begin the names of the age groups in cell A2. In cell B1, enter the year label 2000. In cell C1, enter the year label 2010. Enter the appropriate percentages for the selected age groups by year.

Alternatively, enter decimal values rather than type in percentages. Highlight the cells containing the decimals. In the Home tab, select Cells, Format. On the menu that appears, select Format Cells…. Under the Number tab, select Percentage and input the number of decimal places the values should reflect. Click OK. All values will display as a percent with the designated number of decimal places.

Highlight the block of cells from A1 through cell C5 entered from Table 3.4. Select the Insert tab at the top of the Excel window. Select Pie and click the 2-D icon to make a pie chart.

Answer

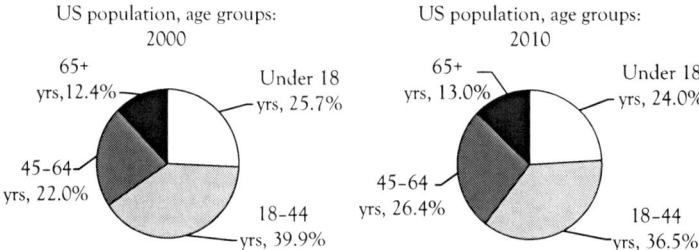

Figure 3.4. Pie charts: US population by selected age groups, 2000 and 2010 (US Census Bureau).[7]

Again you will probably want to improve the basic graph. To insert a chart title, select Layout and highlight Chart Title. Type in the chosen title for the chart. Select Data Labels and click More Data Label Options. Select Category Name, Value, and Show Leader Lines. Click OK.

Discussion

Although the overall population of the United States has grown from 2000 to 2010, which is not shown in the pie charts, the comparative growth rates within the selected age groups differ significantly. The growth rate in the under 18 years category has dropped, as has the growth rate in the category aged 18 to 44 years. This contrasts significantly with the growth rates witnessed in the categories aged 45 to 64 and 65 and older. The population of the United States is aging. Continued growth in the category of 65 years and older can be expected given the increase in the category aged 45 to 64 years. Those service sectors that depend on younger cohorts to fund and focus their expenditures on older Americans are most affected.

3.3.2   Working with Quantitative Data

The tools we used in the previous section can be used to graphically display quantitative data for relatively small samples.

Example 3.6: Forming a Column Chart with Quantitative Data

The Question

According to the Bureau of Transportation Statistics, the total baggage fees collected by airline for the first quarter of 2010, 2011, and 2012 are shown in Table 3.5. Use the data to form and interpret a side-by-side column chart.

Table 3.5. Total Baggage Fees, First Quarter, 2010–2012[8]

	A	B	C	D
	First Quart Baggage Fees by Airline			
1		**2010**	**2011**	**2012**
2	Delta	$217,773	$197,971	$198,352
3	American	$128,539	$137,210	$139,239
4	US Airways	$120,720	$120,925	$124,333
5	United	$71,145	$66,245	$156,761
6	AirTran	$35,005	$39,267	$34,507
7	Alaska	$21,166	$36,201	$33,165
8	Spirit	$16,033	$28,226	$38,023
9	Allegiant	$14,826	$14,462	$16,069
10	Frontier	$13,872	$16,681	$17,201
11	JetBlue	$13,763	$14,278	$17,203
12	Hawaiian	$11,672	$13,498	$16,114
13	Continental	$76,603	$76,304	

Answer

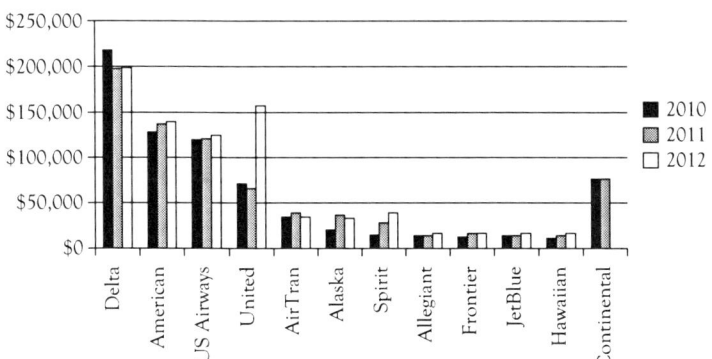

Figure 3.5. A side-by-side column chart: Total baggage fees, first quarter, 2010–2012.[9]

Using Excel

Enter the data as shown in Table 3.5, leaving cell A1 blank. Begin the names of the airlines in cell A2. In cell B1, enter the year label 2010. In cell C1, enter the year label 2011. And in cell D1, enter the year label 2012. Enter the appropriate baggage fees for each airline and year.

Highlight the block of cells from A1 through cell D13. Select the Insert tab at the top of the Excel window. Select Column and click 2-D Column, left-most icon for a side-by-side column chart.

Again you will probably want to improve the basic graph. To insert a chart title, select Layout and highlight Chart Title. Type in the chosen title for the chart.

Discussion

Continental Airlines ceased to exist in 2012, so it was listed last among the airlines because of its unique circumstance. The most striking pattern shown in the side-by-side column chart is the increase in baggage fees for United Airlines during the first quarter of 2012. However, the two events are related: Continental Airlines merged with United Airlines in late 2011, which probably accounts for the increase in total baggage fees United reported during the first quarter of 2012. Baggage fees converted to a per passenger basis will most likely show a different pattern.

The side-by-side column chart above allows the reader to examine the patterns displayed over time across individual airlines. When the patterns across individual years are less important than the combined pattern exhibited over the entire period of time, a stacked column chart is more appropriate.

Example 3.7: Forming a Stacked Column Chart with Quantitative Data

The Question

According to the Bureau of Transportation Statistics, the total baggage fees collected by airline for the first quarter of 2010, 2011, and 2012 are shown in Table 3.5. Use the data to form and interpret a stacked column chart.

Answer

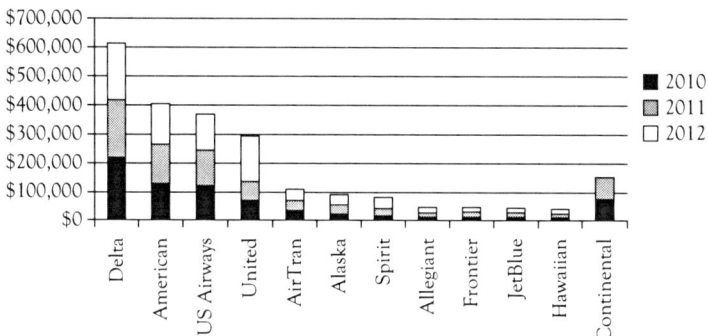

Figure 3.6. A stacked column chart: Total baggage fees, first quarter, 2010–2012.[10]

Using Excel

Highlight the block of cells from A1 through cell D13 entered from Table 3.5. Select the Insert tab at the top of the Excel window. Select Column and click 2-D Column, middle icon to make a stacked column chart.

Again you will probably want to improve the basic graph. To insert a chart title, select Layout and highlight Chart Title. Type in the chosen title for the chart.

Discussion

Collectively the pattern for Delta Airlines shows it has charged more than other airlines. Were baggage fees converted to a per passenger basis, most likely a different pattern would emerge. Had the merger of Continental and United Airlines been reflected in the graph, again the emerging pattern might have been different.

Summarizing data for continuous variables can be achieved in a histogram as seen earlier in this chapter or a variety of other charts, including line graphs and pie charts.

Example 3.8: Forming a Line Graph

The Question

Of interest to any airline passenger is the on-time performance of the airlines traveled. Airlines have centrally reported their year-to-date on-time flight performances for all years since 1995. Use the data in Table 3.6 to develop and interpret a line graph of the on-time performance of airlines from 1995 to 2011.

Table 3.6. Summary of Airline On-Time Performance, 1995–2011[11]

	A	B	C	D
1	1995	79%	2004	79%
2	1996	74%	2005	79%
3	1997	77%	2006	77%
4	1998	77%	2007	74%
5	1999	75%	2008	74%
6	2000	75%	2009	79%
7	2001	77%	2010	80%
8	2002	82%	2011	77%
9	2003	83%		

Answer

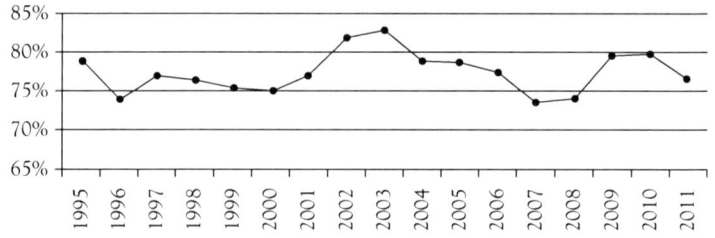

Figure 3.7. A line graph: Percent of on-time airline arrivals, 1995–2011.[12]

Using Excel

Enter year labels 1995 through 2011 in cells A1 through A17 with their related percent of on-time arrivals in cells B1 through B17.

Highlight the block of cells A1 through B17 and click the Insert tab at the top of the Excel spreadsheet. Select Line, and highlight the left-most icon in the second row, Line with Markers. Left click on the resulting graph and highlight Select Data. In the field of Chart Data Range, scroll over the cells B1 through B17. Under the title Horizontal Axis Labels, click Edit. In the field Axis Label Range, scroll over cells A1 through A17. Delete the legend to maximize the graphic display area.

Again you will probably want to improve the basic graph. To insert a chart title, select Layout and highlight Chart Title. Type in the chosen title for the chart. You may want to individualize the styles and colors for the line and the point markers, as well as the font sizes for the labels and the chart title.

Discussion

On-time arrivals peaked in 2003 but matched their all-time low levels in 1996, in 2007, and again in 2008. Undoubtedly increased security procedures instituted in 2009 after the horrific terrorist attacks in the United States have influenced overall US airline performances including on-time arrivals as well as related on-time departures since the last quarter of 2009.

Quantitative data can also be expressed in a line graph for a cumulative relative frequency distribution. An ogive is a special line graph. A "less than or equal to" ogive is a line graph that captures the percent of all data values in the set that fall at or below the value represented on the horizontal axis. It is useful in answering questions such as: At or below what value do 80% of the data values fall?

Example 3.9: Forming an Ogive

The Question

Returning to the highway mileage ratings for a sample of 67 model 2011 vehicles shown in Table 3.2, form and interpret a "less than" ogive to graphically summarize the data, where the upper class limits are used as labels.

Answer

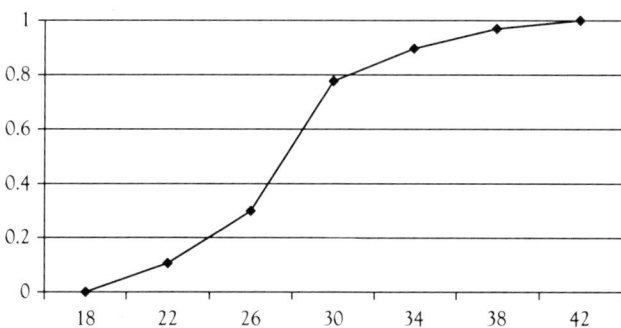

Figure 3.8. A "less than" Ogive: Highway mileage for 2011 model US vehicles.[13]

Using Excel

Return to the spreadsheet on which you entered data as shown in Table 3.2. Copy the class labels from cells C2 through C7 and paste them into cells C15 through C20.

Table 3.7. Cumulative Frequency Distributions

	A	B	C	D
1	*Upper limits*	*Frequency*	Cum freq	Cum relfreq
2	18			0
3	22	7	7	0.104477612
4	26	13	20	0.298507463
5	30	32	52	0.776119403
6	34	8	60	0.895522388
7	38	5	65	0.970149254
8	42	2	67	1

Since ogives rise from the horizontal axis, we need to create the interval before our registered mileage readings to show the interval in which there were no mileages at or below that class. So, in cell C14, add a new class label: 14 to 18. In cell D14, type the value 0. In cell D15, enter the equation: =D14+D2. Drag the equation down over cells D16 through D20. That creates a column of the running total of frequencies registered at or below each class. In cell E14, enter the equation: =D14/67. Drag the equation down over cells E15 through E20. That creates a column of cumulative relative frequencies for each of the classes.

Enter the upper class limit of each of the class intervals into cell C24 through cell C30. You should have the label values: 18, 22, 26, 30, 34, 38, and 42. Copy cell E15 through cell E20 and paste values into cell D24 through cell D30.

Highlight the block of cells from C24 through cell D30. Select the Insert tab at the top of the Excel window. Select Line and click the left-most 2-D icon to make a line chart.

Again you will probably want to improve the basic graph. To insert a chart title, select Layout and highlight Chart Title. Type in the chosen title for the chart. Highlight and delete the legend on the right side of the graph to expand the usable display area. Move the four-way arrow cursor toward the vertical axis until it becomes a single arrow; click your left mouse button. In the Axis Options window that opens, select Maximum: Fixed and enter 1.0 to have the vertical axis maximum set at 1.0. Select Major Unit: Fixed and enter 0.2 to retain the horizontal grid lines at 0.2, 0.4, 0.6, 0.8 and 1.0. At the same time, you can individualize the styles and colors for the line and the point markers before you click OK.

Discussion

Comparing Figure 3.1 with Figure 3.8 demonstrates the information each depicts well. Note that the shallow line segment between 18 and 22 mpg indicates comparatively few vehicles were added by the class 18 up to 22. In contrast, the steep line segment between 26 and 30 mpg indicates many vehicles were added by the class 26 up to 30 mpg. In fact, the cumulative relative frequency jumped from around 25% to nearly 80%, indicating the class 26 up to 30 mpg is the modal class and represented nearly 50% of the values in the data set.

3.3.3 *Working with Bivariate Quantitative Data*

We often find it telling to collect data that track two quantitative variables simultaneously. We may want to look at changes in one variable as it changes with or perhaps even influences values of a second variable. Sometimes a relationship between two numerical random variables becomes apparent by collecting a random sample of measurements of both variables and looking at an *x-y* scatter diagram of the data. Seldom do we see a relationship between two variables so tightly connected that the scatter diagram maps to a straight line. More likely, the data points are somewhat dispersed, or scattered, around the grid. A cloud of data points that generally rises from left to right provides evidence of a *positive* or *direct* relationship between the two variables, whereas a cloud of points that falls from left to right provides evidence of a *negative* or *inverse* relationship between *x*, the independent variable, and *y*, the dependent variable. A cloud of points that appears as a smattering of points with no direction to them provides evidence that the value of *y* is unrelated to the value of *x*. Scatterplots are most compelling when the independent variable is cast as the cause and the dependent variable as its effect. Echoes of the dependent variable as a *function of x* take on new meaning when depicted in an *x-y*, an *independent-dependent*, a *cause-effect* scatterplot.

Example 3.10: Forming a Scatterplot for Two Quantitative Variables

The Question

Is there a relationship between the annual average 6-month Treasury Bill rates and the inflation rate based on the Consumer Price Index (CPI) for urban Americans? To develop a preliminary sense of an answer, we gathered the 27 observations from 1982 through 2008, as shown in Table 3.8. Use these data to develop and interpret a scatterplot, using the Treasury Bill rate as the independent variable and the inflation based on the CPI as the dependent variable.

Table 3.8. *Annual Average 6-Month Treasury Bill Rates with Annual Inflation Based on the CPI*

	A	B	C	D	E	F
1		T-Bill rate	CPI inflation		T-Bill rate	CPI inflation
2	1982	11.06	6.09	1996	5.08	2.82
3	1983	8.74	3.19	1997	5.18	2.41
4	1984	9.78	4.37	1998	4.83	1.63
5	1985	7.65	3.56	1999	4.75	2.17
6	1986	6.02	1.64	2000	5.90	3.27
7	1987	6.03	3.75	2001	3.34	2.83
8	1988	6.91	4.02	2002	1.68	1.62
9	1989	8.03	4.83	2003	1.05	2.39
10	1990	7.46	5.35	2004	1.58	2.65
11	1991	5.44	4.36	2005	3.39	3.32
12	1992	3.54	2.95	2006	4.81	3.23
13	1993	3.12	3.03	2007	4.44	2.90
14	1994	4.64	2.52	2008	1.62	3.78
15	1995	5.56	2.90			

Using Excel

Enter the years 1982 through 2008 in cell A1 through cell A27. Enter the Treasury Bill rate and the inflation based on the CPI in columns B and C, respectively, for each of the years shown in Table 3.8. Note that the data for each variable needs to be listed in a single column, not split into two columns as shown in Table 3.8.

Highlight the block of cells from B1 through C27. Do not include the year in which each of the variables occurred. Select the Insert tab at the top of the Excel window. Select Scatter and highlight the first row, left-most icon, Scatter with Only Markers. Delete the legend to maximize the display area. Select the Legend tab at the top of the screen. Select Axis Titles, and enter the chosen labels for each of the horizontal and vertical axes.

Again you will probably want to individualize the basic graph, controlling fonts, colors, and styles contained in it.

Answer

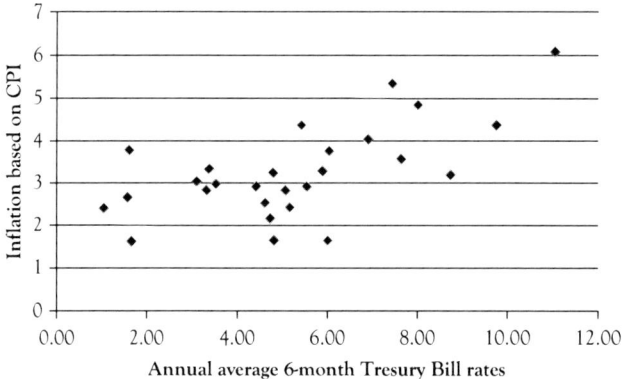

Figure 3.9. A scatterplot: Annual average 6-month Treasury Bill rates with annual inflation based on CPI.

Discussion

In this problem, the annual average 6-month Treasury Bill rate is being used to predict the rate of inflation based on the CPI. So the annual average 6-month Treasury Bill rate is the independent variable, or the cause, on the *x*-axis, and the rate of inflation based on the CPI is the dependent variable, or the effect, on the *y*-axis. The data cloud in Figure 3.9 seems to be rising left to right, indicating there is a positive, or direct, relationship between the annual average 6-month Treasury Bill rate and the annual rate of inflation as based on the CPI. If we knew nothing about the average 6-month Treasury Bill rate, we would guess the average inflation rate to be a little over 3%. But if we knew the annual average 6-month Treasury Bill rate was 10%, we would guess the annual rate of inflation to be around 4.5%. This example is developed more completely in Chapter 8 of the companion publication, *Working with Sample Data: Exploration and Inference.*

CHAPTER 4

Working with Mathematical Models

Most of this book focuses on fundamental mathematical concepts and how to employ those concepts using Microsoft Excel. The practical application of mathematics to an actual decision or analysis involves several related computations. These computations must be carefully designed to be valid and useful. The conceptual framework for these related computations is called a *mathematical model*. One of the primary strengths of Microsoft Excel is the ability to quickly implement powerful mathematical models. When mathematical models are built in Excel, they are often called spreadsheet models. This chapter illustrates some mathematical models created using Microsoft Excel.

4.1   What is a Mathematical Model?

A mathematical model is a simplified representation of a system, either one that existed in the past or may exist in the future. Characteristics of the system are assessed in terms of *variables*. These assessments may be *quantitative*, in terms of measurements or counts, or *qualitative*, in terms of a category or whether a logical condition is true or false. When a model is defined in Excel, a variable is typically a value displayed in a cell.

Mathematical models generally include at least two variables, but often considerably many more. The values of some variables are provided as inputs and the values of other variables are determined by the model. The creation of a mathematical model entails both the definition of its variables and *relationships* between those variables. Relationships between quantitative variables often take the form of algebraic equations, while relationships involving qualitative variables are often logical, if–then rules. In Excel, these relationships are usually in the form of formulas that determine the value of a variable appearing in a cell.

Table 4.1. Select Fahrenheit Temperatures and Equivalent Centigrade Temperatures

	A	B
1	**Fahrenheit temperature**	**Centigrade temperature**
2	-20	-28.9
3	-10	-23.3
4	0	-17.8
5	10	-12.2
6	20	-6.7
7	30	-1.1
8	40	4.4
9	50	10.0
10	60	15.6
11	70	21.1
12	80	26.7
13	90	32.2
14	100	37.8
15	110	43.3
16	120	48.9

An example of a simple model would be a model that converts Fahrenheit temperatures to centigrade. In Excel, the Fahrenheit temperature might appear as an input variable in cell A1 and the centigrade temperature as a variable in cell B1. In cell B1, the formula: $=(5/9)*(A1-32)$ would be entered, and it defines the relationship between the two variables. The user of the model can enter any Fahrenheit temperature in cell A1 and will immediately see the corresponding centigrade temperature in cell B1.

To better understand the relationship between Fahrenheit temperature and centigrade temperature, it would be nice to see how centigrade temperature changes in response to changes in Fahrenheit temperature. While this can be explored by changing the value in cell A1, a clearer picture would come from copying the formula in cell B1 to subsequent cells in column B and entering a series of different Fahrenheit temperatures in column A to create a table, as illustrated in Table 4.1. The table can be represented visually in a chart as shown in Figure 4.1. The ability to expand mathematical models into tables and display them in charts is a key reason why Excel is such a popular tool.

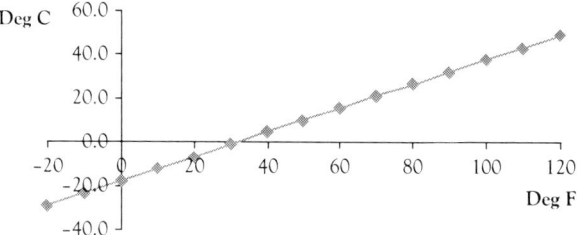

Figure 4.1. Graph of the relationship between Fahrenheit and centigrade temperatures.

Models can serve different purposes. A *descriptive model* endeavors to characterize the operation of an existing system in order to get an improved understanding of it. A *predictive model* seeks to forecast how a system will operate in the future. A *normative model* tries to determine how a system could best operate in terms of certain performance criteria. The type of model might impact how the model variables, the designation of input variables (for which values are provided by the user), and relationships are designed. However, in Excel it is often possible to create a model and then use it for different purposes by applying some of the solution tools built into Excel.

The remainder of this chapter will provide some examples of mathematical models for a fictitious venture of a summer business selling ice cream bars at a beach resort. In each case, the models will be presented as Excel spreadsheet models.

4.2   A Cost Model and Breakeven Analysis

Suppose three college students want to consider operating a summer business selling ice cream bars at a popular beach site. In order to make this decision, they need to have some understanding of the costs involved.

The students recognize that the total cost will depend on how many ice cream bars they sell, but some components of the total cost are directly related to the number of bars sold while other costs are largely unaffected by the sales volume. If the students wanted to double the number of ice bars they sold, they would need to purchase twice as many bars from their supplier, probably have roughly twice the cost in transporting the ice cream bars from the supplier, and probably about double the electricity cost of keeping the items in a freezer. Costs that change with sales volume are called *variable costs*. Someone who sold ice cream bars last year, incurred costs of $10,800 for ice cream bars, transportation, and electricity. This vendor sold 36,000 ice cream bars, so the variable cost per unit was $0.30, or =10800/36000.

The costs that are relatively invariant include the cost to rent a small stand, insurance, and city vendor fees. Also, the students need to recoup enough to compensate for not taking another summer job, so they include an "opportunity cost" equivalent to what they would earn elsewhere. These cost categories that are assumed to be the same regardless of the number of ice cream bars sold are called *fixed costs*. The students decide they will need to cover a total of $40,000 in fixed costs over the summer months when they operate this business.

Based on these estimates, we can develop a mathematical model called a linear cost model. This simple model determines the total cost as the sum of the fixed cost and the variable cost and treats the variable cost as $0.30 times the number of ice cream bars sold. Algebraically, if we let Q represent the number of ice cream bars that the students will sell during the summer and let C represent the total cost, the relationship between these variables is:

$$C = \$40,000 + \$0.30Q$$

We can present this relationship as a table and a chart in Excel. Table 4.2 shows a table prepared in Excel that treats the number of ice cream bars sold as an input variable and shows the corresponding fixed cost, variable cost, and total cost for total unit sales from 0 to 70,000 in increments of 10,000. The fixed cost values in the second column remain at $40,000, while the value in the third column is generated by a formula multiplying $0.30 times the value in the first column for the same row. The total cost is the sum of the values in the second and third columns. The table values are used to prepare the chart in Figure 4.2. From the table and chart, we see that the fixed cost will be the dominant cost component if fewer than 70,000 ice cream bars are sold.

Table 4.2. Fixed, Variable, and Total Cost for Select
Production Quantities for Ice Cream Bar Venture

	A	B	C	D
1	Quantity	Fixed Cost	Variable Cost	Total Cost
2	0	$40,000	$0	$40,000
3	5,000	$40,000	$1,500	$41,500
4	10,000	$40,000	$3,000	$43,000
5	15,000	$40,000	$4,500	$44,500
6	20,000	$40,000	$6,000	$46,000
7	25,000	$40,000	$7,500	$47,500
8	30,000	$40,000	$9,000	$49,000
9	35,000	$40,000	$10,500	$50,500
10	40,000	$40,000	$12,000	$52,000
11	45,000	$40,000	$13,500	$53,500
12	50,000	$40,000	$15,000	$55,000
13	55,000	$40,000	$16,500	$56,500
14	60,000	$40,000	$18,000	$58,000
15	65,000	$40,000	$19,500	$59,500
16	70,000	$40,000	$21,000	$61,000

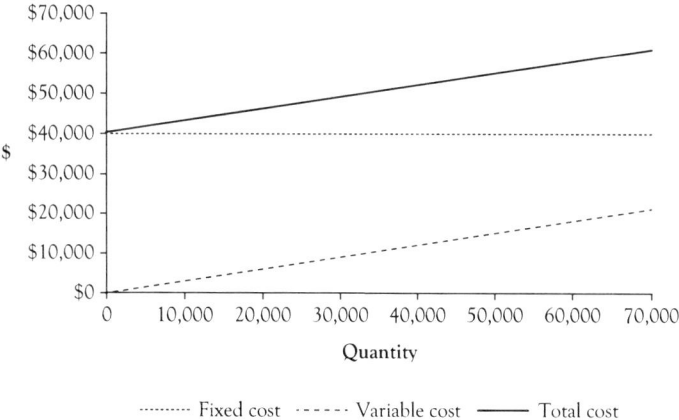

Figure 4.2. Graph of cost components for ice cream bar venture.

In order for this venture to be worthwhile, the students will need to collect enough from sales of the ice cream bars to cover their costs. The total receipts from sales of the ice cream bars will be the *revenue* of their venture. They will need to decide on a price to charge for the ice cream

Table 4.3. *Cost, Revenue, and Profit for Select Production Quantities for Ice Cream Bar Venture at Price of $1.50*

	A	B	C	D
1	Quantity	Cost	Revenue	Profit
2	0	$40,000	$0	-$40,000
3	5,000	$41,500	$7,500	-$34,000
4	10,000	$43,000	$15,000	-$28,000
5	15,000	$44,500	$22,500	-$22,000
6	20,000	$46,000	$30,000	-$16,000
7	25,000	$47,500	$37,500	-$10,000
8	30,000	$49,000	$45,000	-$4,000
9	35,000	$50,500	$52,500	$2,000
10	40,000	$52,000	$60,000	$8,000
11	45,000	$53,500	$67,500	$14,000
12	50,000	$55,000	$75,000	$20,000
13	55,000	$56,500	$82,500	$26,000
14	60,000	$58,000	$90,000	$32,000
15	65,000	$59,500	$97,500	$38,000
16	70,000	$61,000	$105,000	$44,000

bars and then hopefully sell enough bars so that the revenue will be higher than the total cost. If the revenue is higher than the total cost they will earn a positive *profit*. However, if the revenue is less than the total cost, they will have a negative profit, or *loss*.

Suppose the students decided to charge $1.50 for each ice cream bar. The revenue, symbolized by variable R, would be a function of the quantity Q sold:

$$R = \$1.50Q$$

The profit from the operation, symbolized by variable Π (*pi*), would be determined from the difference of revenue and cost:

$$\Pi = R - C$$

Table 4.3 displays a table of revenue, cost, and profit associated with selected quantities. Figure 4.3 shows a chart prepared from the table.

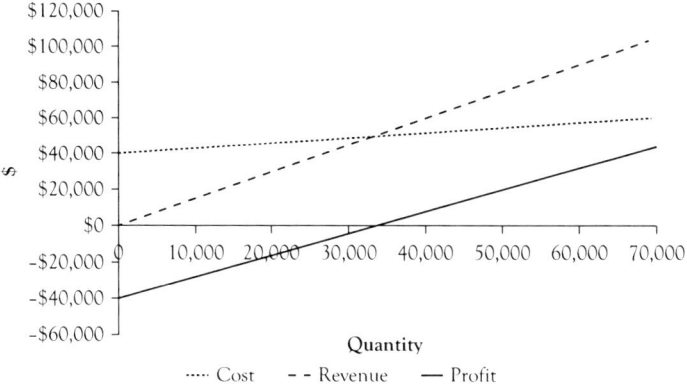

Figure 4.3. Graph of cost, revenue, and profit functions for ice cream bar venture at price of $1.50.

Figure 4.3 indicates that the students will earn a positive profit if they can sell more than 35,000 ice cream bars during the summer. The actual quantity where the result changes from a loss to a profit is known as the *breakeven point*. A precise value for the breakeven point can be determined in a number of ways. One technique is recognize that each ice cream bar sold results a unit profit contribution represented by the difference between the price charged and the variable cost per unit and the venture needs to accumulate enough unit contributions to offset the fixed cost. So the breakeven quantity in this case would be:

$$Q_{BE} = \frac{\$40,000}{(\$1.50 - \$0.30)} = 33,333.3 \text{ units}$$

So if the students sell 33,334 ice cream bars or more, they have a profit. Note that the profit line shown in Figure 4.3 crosses the *x*-axis into positive values at the breakeven point.

Another approach to finding the breakeven point is to consider that the price needs to offset the average cost of an ice cream bar when total cost is divided by the quantity of units sold. Table 4.4 shows the computation of the average cost per ice cream bar at different values of the quantity Q. Figure 4.4 shows a chart with a graph of the average cost with a horizontal line at the price charged of $1.50. Note that the average cost curve drops below $1.50 at 33,334 units.

Table 4.4. Average Cost Per Unit for Select Production Quantities for Ice Cream Bar Venture

	A	B	C
1	Quantity	Total Cost	Average Cost
2	0	$40,000	
3	5,000	$41,500	$8.30
4	10,000	$43,000	$4.30
5	15,000	$44,500	$2.97
6	20,000	$46,000	$2.30
7	25,000	$47,500	$1.90
8	30,000	$49,000	$1.63
9	35,000	$50,500	$1.44
10	40,000	$52,000	$1.30
11	45,000	$53,500	$1.19
12	50,000	$55,000	$1.10
13	55,000	$56,500	$1.03
14	60,000	$58,000	$0.97
15	65,000	$59,500	$0.92
16	70,000	$61,000	$0.87

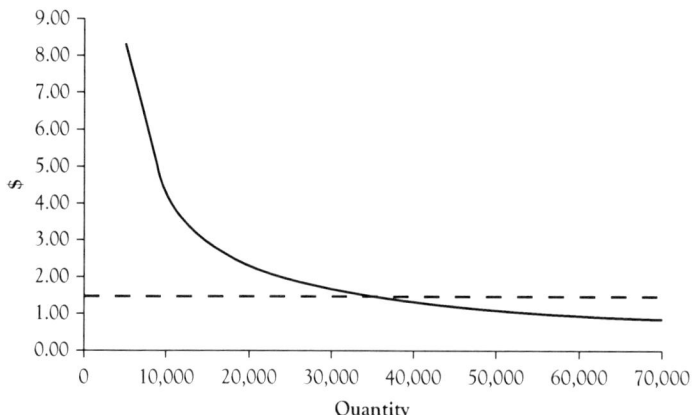

Figure 4.4. Graph of average cost curve for ice cream bar venture in comparison to price of $1.50 per ice cream bar.

While formulas to calculate the breakeven volume can be entered into a worksheet cell to display the value, another approach is to use the Goal Seek feature of Excel to find a value of Q that makes the profit exactly zero. To use Goal Seek, we place a tentative value for the breakeven point in a cell and then enter a formula to calculate the profit associated with that quantity in a neighboring cell. Suppose we use cells I2 and J2 for this illustration. Enter 30,000 as a test value in cell I2 and the formula: =1.5*I2–40000–0.3*I2 into cell J1. Initially you will see the value –4000 in cell J2, indicating that the students would lose $4,000 if they only sold 30,000 ice cream bars.

Next, we access the Goal Seek capability built into Excel. In Excel 2007 and Excel 2010, this feature is accessed by clicking "What-If Analysis" on the Data ribbon and then selecting "Goal Seek...." (On earlier versions of Excel, the Goal Seek module is accessed from the Tools menu.) An input form appears requiring three entries. In the "Set Cell" box, we enter J2 as the cell value we want the Goal Seek operation to change as specified. In the "To Value" textbox we enter a zero to indicate that we want the value in J2 to be changed to a zero. In the "By Changing Cell" textbox, we enter I2 as the cell we want Goal Seek to modify to achieve a zero profit. Entering these values and clicking OK will change the value in cell I2 to 33333.33 and the value in cell J2 to zero.

The model can easily be changed to examine the impact of charging a different price. Suppose we wanted to consider charging $2.00 per ice cream bar. Simply changing the prices in column E to $2.00 and looking at the graphs, we see the breakeven quantity drops below 25,000 units. Changing the formula in cell J2 to: =2*I2–40000–0.3*I2 and repeating a Goal Seek operation shows the breakeven quantity is 23,529.4 units.

By exploring other prices, the reader will quickly discover that the higher the price charged, the lower the breakeven sales quantity and that it is possible to compute a breakeven quantity for any price that is higher than $0.30. (However, for prices close to $0.30, the breakeven quantity will be very high!) The ease and speed with which models can be changed is one reason why Excel is such a popular platform for mathematical modeling.

4.3   The Ice Cream Bar Venture Model with a Demand Curve

The analysis in the previous text section determined the sales volume necessary to break even for any given price. However, there is no guarantee that there will be enough customer purchases to reach the breakeven level, and while a higher price will lower the breakeven level, a higher price will also discourage more potential customers from making a purchase. To make our model more realistic, we need to consider how price will affect our potential sales. The standard means of doing that in a mathematical model is with a *demand curve*.

A demand curve is a relationship between the price that is charged and the maximum quantity of units that would be sold at that price. The convention in economics is to draw demand curves with quantity on the horizontal axis and price on the vertical axis, so for any specified purchase quantity, the demand curve indicates the maximum price the seller can expect to charge and sell that many units. The *law of demand* in economics states that increases in price will result in decreases in the quantity sold, and vice versa, so demand curves are inversely related in general and negatively sloped when graphed.

One means of developing a demand curve is to estimate the maximum prices that can be charged for selected quantities and then infer a relationship between quantity and maximum possible price that fits those estimates. One simple demand curve is a linear demand curve, which is a straight line on a graph, and requires estimates of only two quantity–price pairs. Based on sales in the prior summer season, the students expect they would sell about 36,000 ice cream bars if they charged a price of $1.50. However, if they raised the price to $2.00 they estimate that the total sales for the summer would drop to 26,000 ice cream bars.

Figure 4.5 shows a graph of the linear demand curve that is consistent with the above quantity–price estimates. Using basic algebra, the following algebraic formula can be derived for this linear demand curve:

$$P = \$3.3 - \$0.00005Q$$

where Q is quantity of ice cream bars that the students consider planning to sell and P is maximum price they could charge to sell that many ice cream bars.

With the demand curve, the students can explore what price they should charge, accounting for the reality that a higher price will reduce the potential sales. Recalling from the previous section that the breakeven quantity was 33,334 units at a price of $1.50 and the breakeven

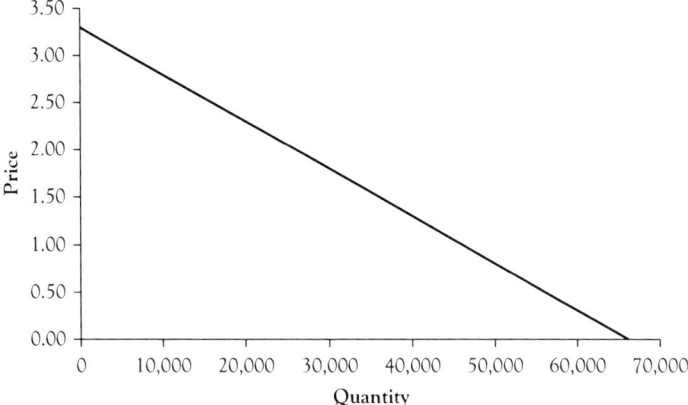

Figure 4.5. Graph of linear demand curve for ice cream bar venture.

quantity at a price of $2.00 was 23,530 units, we see that students can make a profit at either price. However, the expected profits if they sell the amounts estimated from the demand curve will differ. If the students sell 36,000 units at a price of $1.50, the revenue will be $54,000 and the total cost will be $50,800, resulting in a profit of $3,200. If the students sell 26,000 units at a price of $2.00, the revenue will be $52,000 and the total cost will be $47,800, resulting in a profit of $4,200.

We can explore the impact of the demand curve on the revenue, cost, and profit for all possible choices of quantity volume by replacing the price in the revenue function with the relation that links maximum price to quantity from the demand curve equation. The revenue function becomes

$$R = \$3.3Q - \$0.00005Q^2$$

since revenue = price * quantity, or $R = P{*}Q = Q(\$3.3 - \$0.00005Q)$. The cost function is not affected by the demand curve and remains:

$$C = \$40,000 + \$0.30Q$$

The profit function, which is the difference between revenue and cost becomes:

$$\Pi = -\$40,000 + \$3.00Q - 0.00005Q^2$$

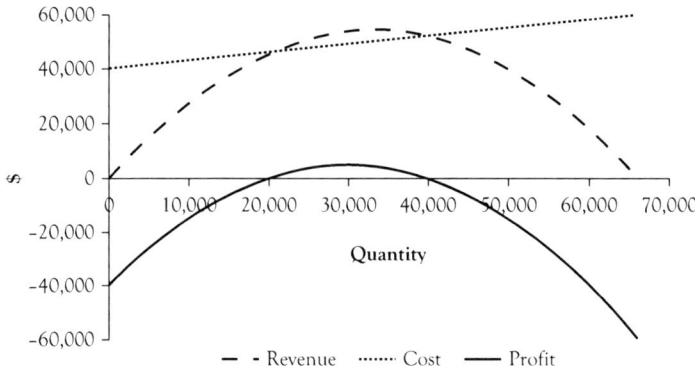

Figure 4.6. Graph of cost, revenue, and profit functions assuming price associated with quantity on linear demand curve.

Figure 4.6 shows a graph of these three functions for the ice cream bar venture. The graph indicates there is an interval of sales amounts where a profit can be expected if the maximum price is charged base on the demand curve. The maximum appears to be at a volume of approximately 30,000 units. At the same time, it is possible for the venture to have a loss if the volume is too high or too low. At volumes below 20,000 units, despite the fact that a higher price can be charged, there are not enough sales to offset the $40,000 fixed cost. To sell volumes over 40,000 units, the price must be reduced because the margin between the maximum price that can be charged and the variable cost per unit of $0.30 per ice cream bar is not sufficient to offset the fixed cost.

We can determine the best quantity–price pair from the algebraic equations in the model. Since the profit function is a quadratic function, the maximum will occur at the vertex of the quadratic equation. For a quadratic equation of the form

$$y = ax^2 + bx + c$$

the vertex occurs at $x = -b/2a$. For the profit function, the x variable is the quantity Q, the coefficient $a = -0.00005$, and $b = 3$. So, for this function, the vertex where profit reaches its maximum is at the quantity of

$$Q = \frac{-3}{2(-0.00005)} = 30,000 \text{ units}$$

Using the demand curve, the maximum price that can be charged to sell 30,000 ice cream bars during the summer is $1.80. If you set Q to 30,000 in the profit function, the resulting profit will be $5000.

For Windows users, Excel provides a feature to find the quantity in the mathematical model that maximizes the profit called Solver. Like the data analysis toolpak described in Chapter 3, the Solver comes as part of Excel for Windows, but needs to be installed as an add-in. In Excel 2007 and later, you can activate Solver by clicking the File tab, then clicking "Add-Ins" on the left menu. Next, select "Excel Add-ins:" in the drop-down list box at the bottom of the window next to the label "Manage," and click GO. A new window of Add-ins will appear. Check the box next to "Solver Add-in" and click OK. (The add-in is available for earlier versions of Excel, but needs to be activated differently.)

To use the Solver to find the best quantity–price pair on the demand curve, we place our guess for the quantity in one cell and a formula with the profit associated with that quantity in another cell. Suppose we enter 36,000 in cell A1 and the formula for profit: =−40000+3*A1−0.00005*A1*A1 in cell B1. In Excel 2007 and Excel 2010, go to the Data ribbon. Solver should be available in the analysis group on the ribbon. Click "Solver" and a form appears. In the box next to "Set objective," enter "B1." On the next line, click the radio button "Max." Under "By Changing Variable Cells:," enter A1, and click the common button labeled "Solve." A window "Solver Results" appears. Click OK. If you look at cells A1 and B1, you will see that cell A1 has been changed to 30,000 and cell B1 shows the profit associated with a quantity of 30,000 ice cream bars of $5000.

As we noted in the first section of this chapter, models are simplified representations of a real situation. When the real situation is an estimate of what will occur in the future, such as the ice cream venture, the actual outcome is almost certain to differ from what a model will estimate. The students will decide on a price and find out later how many ice cream bars they will sell. If they charge $1.80, the model estimates they will sell 30,000 ice cream bars and earn a profit of $5,000, but what if the model is wrong?

The students will be pleased if they can sell more than 30,000 units, as this will increase their profit by $1.50 for each additional ice cream bar they sell. However, if the sales fall short of 30,000 units, they might suffer a loss.

The breakeven analysis in the previous section provides a means of assessing how much room for error they have and still remain profitable. Repeating the analysis at a price of $1.80 per ice cream bar shows the breakeven point to be only 26,667 units. So, as long as the students can sell at least 89% of the expected volume, they will not lose money. (Recall that the students included in the fixed cost an amount to offset the lost income from not taking a summer internship.)

In the first section of this chapter, we distinguished descriptive models, predictive models, and normative models. Sections 4.2 and 4.3 illustrated a simple example of each. The cost function developed in Section 4.2 was based on known fixed costs and past records of variable costs, so the cost function we developed is really a descriptive model. When we enhanced the model in this section with a demand curve to explore the change in estimated unit sales corresponding to different ice cream bar prices in the upcoming season, we created a predictive model. Finally, when we used the Solver to find the quantity/price point on the demand curve that would generate the highest expected profit, we employed these relationships as a normative model.

The development of useful mathematical models for business decisions requires an understanding of economics, finance, and operations management principles as they apply to a business. For a discussion of the economic reasoning behind the models in Sections 4.2 and 4.3, see the Business Expert Press book, *Managerial Economics: Concepts and Principles*.

4.4   Planning Daily Operations for the Ice Cream Bar Venture

The models in the previous sections looked at sales for the entire summer season. However, these sales need not occur at a uniform rate each day of the season when ice cream bars are sold. In fact, the students learned that sales can vary considerably from day to day, with the daytime temperature being the dominant influencing factor. On warm days, there are generally more people at the beach location, and those who are there are more eager to buy ice cream bars.

In operating the ice cream bar business, it is important for the students to anticipate the volume of sales for an upcoming day. The students

Table 4.5. Record of Daily Sales for Ice Cream Venture in Prior Summer and Associated High Temperature, Sorted by Sales Volume

	A	B	C	D	E	F	G	H
	Unit	High	Unit	High	Unit	High	Unit	High
1	Sales	Temp	Sales	Temp	Sales	Temp	Sales	Temp
2	92	67	291	74	347	79	418	83
3	133	59	292	72	349	75	424	78
4	139	70	295	77	352	70	425	73
5	147	66	295	79	353	80	429	88
6	148	65	298	75	355	76	437	84
7	169	71	305	75	357	80	444	77
8	184	63	305	76	358	84	450	87
9	189	69	306	69	359	79	452	82
10	194	68	307	76	364	79	454	82
11	208	64	312	78	366	81	455	81
12	209	68	313	78	371	72	458	81
13	210	67	317	81	371	74	463	80
14	211	71	321	77	374	78	463	83
15	218	70	326	71	377	73	477	84
16	220	73	326	75	378	73	488	83
17	247	76	331	74	381	79	507	89
18	249	66	335	72	382	86	512	85
19	250	70	335	75	383	87	517	82
20	261	72	336	76	391	78	530	91
21	263	62	337	77	391	88	541	85
22	266	75	337	77	392	79	548	86
23	272	74	340	73	396	80	550	85
24	273	71	342	80	403	81	553	92
25	277	77	342	84	403	82	595	95
26	290	69	346	83	417	74	608	90

will need to have enough ice cream bars on hand for the day's sales, but want to avoid having an excessive amount as this would result in extra work effort to keep the bars frozen and avoid spoilage.

The person who operated the ice cream bar business in the prior summer season kept records of his daily sales of ice cream bars. One of the students decided to explore the relationship between those daily volumes and the high temperature at the beach location that day. Table 4.5 shows a listing of daily sales from last year, ranked from lowest to highest, with the high temperature that day in the adjacent column. An examination of the table shows there is a significant relationship between daily sales and the high daily temperature. While the numbers tended to be an

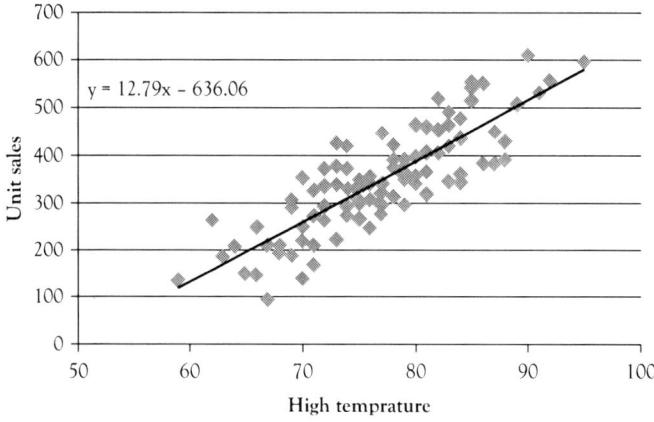

Figure 4.7. Scatterplot of daily sales and associated high temperature in prior summer, with linear trendline and trendline equation.

average of around 350 bars a day, the lowest daily sale volume was below 100 units and the highest sales volume exceeded 600 units. When daily sales were less than 200 ice cream bars, the high temperature was generally under 70 degrees. When daily sales exceeded 500 ice cream bars, the high temperature was typically above 80 degrees.

Figure 4.7 presents these data in the form of a scatterplot. Since the purpose of the analysis is to determine how effective the high temperature was in explaining the daily sales, this scatterplot assigns the high temperature as the cause to the horizontal axis and daily sales as the effect (in numbers of ice cream bars sold) to the vertical axis. While the points in the scatterplot form a cloud, the cloud has a definite positive orientation with the cloud trending higher as the high temperature increases.

The scatterplot in Figure 4.7 has a straight line running through the data. This line is the *linear trendline*, which is the straight line graph that best approximates the relationship between daytime high temperature and ice cream bar sales in the scatterplot. This trendline is "best" in the sense that if you measure the difference between the sales associated with each point on the scatterplot and the sales on the trendline corresponding to that same high temperature, and then square those differences and sum the resulting squared differences, you get the smallest possible total. Thus, this is called a "least-squares fit." To see this trendline for a scatterplot in Excel 2007 and later, select the chart and select the ribbon called "Layout" that is in the "Chart Tools" group. On the ribbon, click the button for "Trendline" and then select "Linear Trendline."

Figure 4.7 also shows an equation for the linear trendline:

$$y = 12.79x - 636.06$$

The variable x is the variable on the horizontal axis (high temperature for the day) and the variable y is the expected daily sales according to the trendline equation. To see this trendline on a scatterplot, select the same "Trendline" button used to get the linear trendline, click the menu option for "More Trendline Options," and then click the radio button for the "Linear" trend/regression type.

The linear trendline equation provides a means of calculating the typical sales last season based on the high temperature that day. If you enter a high temperature in one cell of an Excel worksheet and use the right hand side of the trendline equation as a formula in another cell replacing the x variable with the location of the cell with the temperature value, the value in the cell with the formula will show a typical sales volume at that high temperature. For example, if the high temperature on a summer day in the region was 80 degrees, the typical sales volume was 387 ice cream bars.

The x-coefficient 12.79 in the equation has an interesting interpretation: Ice cream bar sales last year typically increased by about 13 units for each degree increase in the local high temperature. Similarly, an increase of 10 degrees in the high temperature resulted in about 128 more sales of ice cream bars that day, or 10 * 12.79 rounded to the next whole ice cream bar. So, indeed the high temperature has a significant impact on ice cream bar sales.

The trendline equation serves as a descriptive model of typical daily sales last year for a given high temperature. The students would like to have a predictive model that they can use to estimate how many sales to plan for the upcoming day in the next summer season based on a forecast for the high temperature. While it may be reasonable to assume that the relationship between high temperature and daily sales last summer would apply again this summer, based on the analysis in the previous section, the students decided to charge $1.80 per ice cream bar this summer rather than the $1.50 per bar charged last summer. Since the estimated sales for the entire season at a price of $1.50 is 36,000 units, while the estimated sales for the entire season at a price of $1.80 is 30,000, we could make the assumption that a good estimate of daily sales at the higher price of $1.80 on any given day would be 30,000/36,000 or 5/6 of the expected sales at last year's lower price of $1.50. We can reflect this in the linear trendline model, by multiplying both coefficients in the equation by 5/6. The resulting equation would be:

$$y = 10.66x - 530.05$$

At a high temperature of 80 degrees, this model estimates sales would be 323 ice cream bars for the day. Each increase in the high temperature by one degree will typically result in about 11 more unit sales.

The modified model provides a reasonable means of estimating daily sales at the higher price. However, the students realize that even if the model is valid, there will be fluctuation in the actual unit sales from the model forecast. In the case of this venture, it is probably better to over-estimate, rather than underestimate daily sales because there is a greater cost to missing sales than there is to handling excess inventory that are not sold. A sales opportunity is lost if they run out of ice cream bars. Looking at the scatterplot for sales last year, it appears that unit sales on any given day only rarely exceeded the estimate from the trendline by more than 120 units. Adjusting this error margin by the 5/6 reduction due to the increased price this next year, the students might expect that their actual daily sales are unlikely to exceed the estimate of their model by more than 100. So, if the students stock at least 100 more ice cream bars than the predicted unit sales from their model, they are not likely to run out or miss too many sales if they do. They decide to adjust their daily stocking model accordingly as follows.

$$y = 10.66x - 430.05$$

Where x represents the predicted daytime high temperature. So, a day when the high temperature is forecast to be 80 degrees, they will plan to have about 423 ice cream bars in stock at the beginning of the day.

Although high temperature is an effective technique of estimating demand, the students could consider additional variables, such as the day of the week or whether it is a holiday. They could also create models that adjust the forecast for the following day in light of the errors that occurred between actual sales and the forecast on previous days. These enhancements will not be explored here. However, the interested reader can read a book on time series analysis or multiple regression to understand how to create such models.

4.5   A Capital Investment Decision Model

The students operating the ice cream venture leased a small facility for operating their business near the beach. The owner of the facility indicated that he was planning to sell the facility at the conclusion of the summer season and asked the students if any of them were interested in purchasing it. He said he would sell it for $150,000.

One of the students inherited some money and is interested in the possibility of investing in the purchase of the beach property. However, he wants to make sure that the amounts he will get in terms of net income from owning the property and eventually reselling the property are large enough to justify this investment over an investment in stocks, bonds, or certificates of deposit.

The student decides he will develop a model that estimates the net income he will receive if he purchases the building at the conclusion of this summer season, holds it as a lease property for 5 years, and then resells it. For the purpose of the analysis, he assumes that he will be able to resell the building in 5 years for $160,000. The owner of the building shares his financial records from the building and it appears that the owner will realize an income, net of expenses for upkeep, and property taxes of $10,000 this year. The interested student believes he can raise the lease amount at about the rate of the increase of the cost of living, so he will assume his net income would rise $200 each year he owns and leases the property. Table 4.6 shows a portion of the spreadsheet with the assumed net cash flows for the investment, with Year 0 referring to the present year when the student would purchase the facility if he decides to pursue it.

Table 4.6. Projected Cash Flows for 5-Year Investment in Beach Facility

	A	B
1	Year	Net Cash Flow
2	0	$ (150,000.00)
3	1	$ 10,200.00
4	2	$ 10,400.00
5	3	$ 10,600.00
6	4	$ 10,800.00
7	5	$ 171,000.00

Note the projected cash flow for Year 5 reflects the lease income, $11,000, plus the assumed sale price of the property, $160,000. A key consideration to his decision is the alternative rate of return that the student expects he could get if he invested the money elsewhere and the riskiness of alternatives compared to this opportunity. The student realizes he would earn less on certificates of deposit, and would probably earn less on bonds, but those investments involve less risk than owning and leasing the beach property. The student decides that the level of risk he faces owning the beach property is equivalent to the risk of purchasing stocks that would expect to return 9% over the next year. So, he decides he will analyze whether this investment would be better than simply investing the money in an investment that returns 9% every year. However, we will build a model that allows the student to change this required rate of return to see how it affects the decision.

One approach to developing a model is to estimate how much money the student would have if he invested the $150,000 and earned 9% a year compounded annually, and then compare that amount to what the student would have at the end of 5 years after reselling the facility. Since the student would realize some income in the previous 4 years, we will assume those amounts would be reinvested and earn 9% a year.

Table 4.7 shows a spreadsheet model to make this comparison. We use the FV function to determine the amounts that would be realized from alternative investments or reinvestments of income in the earlier years. The formula to determine the future value of the net income received is: =−FV(B$1,5−A6,,B6). Recall the negative sign is needed to distinguish

Table 4.7. *Future Value of Income Flows from Investment in Beach Facility at 9% Rate of Return Compared to Investing $150,000 for 5 Years at 9%, Compounded Annually*

	A	B	C
1	Return Rate:		9.00%
2			
3	Purchase Facility and Hold for 5 Years:		
4			
5	Year		Future Value in Year 5
6	1	$ 10,200.00	$14,398.13
7	2	$ 10,400.00	$13,468.30
8	3	$ 10,600.00	$12,593.86
9	4	$ 10,800.00	$11,772.00
10	5	$ 171,000.00	$171,000.00
11			
12	Total Future Value		$223,232.29
13			
14	Alternative Investment:		
15			
16	Year		
17	0	$ 150,000.00	$230,793.59

cash outflows from cash inflows. Since the formula simply determines the future value of a lump sum for 4 years compounded annually at 9%, there is no payment amount and the third element of the Excel function must be left blank, which is why there is a double comma in the argument.

The FV formula is designed so it can be copied to find the future value of amounts invested or reinvested in future years. Note that the future value of the income in Year 5 and proceeds from the resale of the facility is the same as the actual amount received since there is no time to reinvest these amounts.

Comparing the sum of the future value of the investment of the facility ($223,232.29) to the future value of an alternative investment earning 9% a year ($230,793.59), we see that the student would be better off with the alternative investment in stocks. However, if the rate of return for an alternative investment of equal risk were lower, say 7%, the conclusion

Table 4.8. Net Present Value of Income Flows from Investment in Beach Facility Using 9% Discount Rate, Compared to Upfront Payment of $150,000

	A	B
1	Return Rate:	9.00%
2		
3	Purchase Facility and Hold for 5 Years:	
4		
5	Year	
6	1	$ 10,200.00
7	2	$ 10,400.00
8	3	$ 10,600.00
9	4	$ 10,800.00
10	5	$ 171,000.00
11		
12	NPV	$145,085.67
13		
14	Alternative Investment:	
15		
16	Year	
17	0	$ 150,000.00

would be different and would mean he should change his decision from investing in stocks to purchasing the beach property. Changing the input value in cell A1 to 7%, the facility investment has an accumulated future value of $220,802.51, while the future value of the alternative investment would be only $210,382.76.

Another way of constructing the model is to evaluate the investment alternatives in terms of present value. With this approach we discount any income received in the future to what amount would have to be invested today at the assumed rate of return to have that much at the time it was received. If the sum of these amounts exceed the $150,000 purchase price for the facility, the investment would be worthwhile. Table 4.8 shows the investment decision model using present value. The formula for determining the net present value of the facility investment uses the NPV function: =NPV(B1,B6:B10).

The net present value of the investment in the facility is only $145,085.67. Since this is less than the $150,000 that the student would

Table 4.9. Result of Applying Goal Seek Tool in Excel to Find the Rate of Return for Which the Income Flows from Investment in Beach Facility have a Net Present Value of $150,000

	A	B
1	Return Rate:	8.18%
2		
3	Purchase Facility and Hold for 5 Years:	
4		
5	Year	
6	1	$ 10,200.00
7	2	$ 10,400.00
8	3	$ 10,600.00
9	4	$ 10,800.00
10	5	$ 171,000.00
11		
12	NPV	$150,000.00
13		
14	Alternative Investment:	
15		
16	Year	
17	0	$ 150,000.00

need to invest upfront, this is not a good investment under the assumption of a 9% rate of return. However, if the assumed rate of return for an alternative investment of equivalent risk were only 7%, the net present value of the facility investment would be $157,429.14 and would exceed the $150,000 initial investment.

Both of the models presented here indicate that the facility purchase is worthwhile if the rate of return expected from the investment is sufficiently low. There is a rate of return where the return from the facility investment is exactly the same as for an alternative investment in stocks. From the analysis earlier in this section, we know that rate will be between 7% and 9%. We can determine what rate of return results in an indifferent conclusion by using the Goal Seek capability we used earlier. Remember that Goal Seek is found by clicking the "What-If Analysis" button on the Data ribbon and then clicking the menu entry, "Goal Seek …." Using the net present value model, we ask Excel to set the net present value of the facility investment in cell A13 to 150,000 by changing the rate of return in cell B1. The result of doing this operation shows that the student would be indifferent between the investment alternatives for a rate of return of 8.18%. Table 4.9 shows the result of this operation.

The computation of the rate of return that results in indifference between investing in this project versus investing in another project of equivalent risk is known in finance as the *internal rate of return*. Excel has a built-in IRR function that allows a user to calculate this rate from a formula. The function is discussed in Chapter 1 in Example 1.14.

This chapter presented several models of typical situations faced by a business. From these examples, we can see that Excel is a flexible and powerful tool for developing mathematical models quickly, enabling the user to rapidly explore the effect of changing assumptions in the model. Although all of the situations examined here were fairly simple, Excel can easily handle more complex decision-making scenarios. From these examples, it is easy to see why Excel is the most popular software tool for developing mathematical models. However, caution should be used to reflect the condition that, in order to use Excel effectively, the user needs to have the basic mathematical understanding to use its capabilities properly.

CHAPTER 5

Working with Personal Planning Over Time

We age and, over time, our incomes peak at some point. For those of us who went to work out of high school, incomes tend to peak early. For those of us who went for college or even graduate school, our incomes tend to peak later. For many of us, incomes have already peaked. The salary raise, the increased company profit, the greater rate of return are goals we strive toward but may remain elusive. Market instability, corporate vigor, job security, even personal health, may offset some of the financial advances we have made in the past. Sometimes we have to make do with less because of unforeseen developments. Taking into account various potential scenarios takes on a new level of importance in changing environments.

5.1   Retirement Planning

We plan for our golden years to remain golden, filled with sufficient gold to see us through our lives. Maintaining desired standards of living into the future in stable environments can be challenging. Let us make a few simple assumptions and develop a projection.

Example 5.1: The Present Value of a Stream of Payments into the Future

The Question

A couple estimates they will need $3,000 a month to supplement their Social Security benefits to maintain their current standard of living. If they retire at age 65 and anticipate living to age 90, how much should they have in the bank when they retire if we assume a constant rate of 2% annual growth compounded monthly?

Answer

The present value of a regular stream of payments can be calculated using the equation:

$$P = R \cdot \left(\frac{1-(1+i)^{-n}}{i} \right)$$

where R = the amount of the payment, i = the periodic interest rate, and n = the number of payments made. This is consistent with Example 1.13 in Chapter 1.

Table 5.1. Present Value of Monthly Payments of $R Assuming r% = 0.02 Interest Over t = 25 Years

	A	B
1	Present Value of a Stream of Payments	
2		
3	Payment, R	$3,000
4	Annual interest rate, r	0.02
5	Number of years, t	25
6		
7	Monthly interest rate, i	0.001667
8	Number of interest periods, n	300
9		
10	Current dollars needed to fund	$707,790

Using Excel

The model is designed so the user enters the three values highlighted: the monthly payment, R, entered in cell B3; the annual interest rate, r, entered in cell B4; and the number of years, t, over which the payments should be funded, entered in cell B5. In cell B7, enter the equation: =B4/12. In cell B8, enter the equation: =B5*12. In cell B10, enter the equation: =B3*((1−(1+B7)^−B8)/B7), which is the equation representing the formula for the present value, P. Alternatively, in cell B10, enter the PV equation from the Insert Function: =−PV(B7,B8,B3), using cell referencing. This is consistent with Example 2.15 in Chapter 2. Entering the equation with cell references retains flexibility for the spreadsheet model, allowing the user to input a new value for the payment, the annual interest rate, the number of years, or both, and the current dollars needed to fund the stream of payments will be recalculated.

Discussion

To supplement their Social Security benefits by $3,000 a month begin-
ning at age 65 to be drawn for 25 years to age 90, assuming a constant
rate of 2% annual growth compounded monthly, the couple should have
saved a total of $707,790 by age 65.

Example 5.2: The Future Value of a Stream of Payments Made into the Future

The Question

A couple estimates they will need $3,000 a month to supplement their
Social Security benefits to maintain their desired standard of living. If
they retire at age 65 and anticipate living to age 90 and they start saving
at age 30, how much should they save each month from age 30 to age 65
to have $707,790 required to supplement their Social Security benefits
by $3,000 a month, all assuming a constant rate of 2% annual growth
compounded monthly?

Answer

The future value of a regular stream of payments into the account can be
calculated using the equation:

$$S = R \cdot \left(\frac{(1+i)^n - 1}{i} \right)$$

where R = the amount of the payment, i = the periodic interest rate,
and n = the number of payments made. This is consistent with
Example 1.12 in Chapter 1.

Using Excel

The model is designed so the user enters the three values highlighted:
the desired future amount, S, entered in cell B3; the annual interest
rate, r, entered in cell B4; and the number of years, t, over which the
payments should be funded, entered in cell B5. In cell B7, enter the

Table 5.2. *Future Value of Monthly Payments of $R Assuming r% = 0.02 Interest Over t = 35 Years*

	A	B
1	**Future Value of a Stream of Payments**	
2		
3	Desired Future Amount, *S*	$707,790
4	Annual interest rate, *r*	0.02
5	Number of years, *t*	35
6		
7	Monthly interest rate, *i*	0.001667
8	Number of interest periods, *n*	420
9		
10	Monthly payment needed to fund	$1,165

equation: =B4/12. In cell B8, enter the equation: =B5*12. In cell B10, enter the equation: =B3/(((1+B7)^B8−1)/B7), which is the equation representing the formula for the monthly payment, *R*, needed to generate $S compounding monthly at *i*% over *n* interest periods. Alternatively, in cell B10, enter the PMT equation from the Insert Function: =−PMT(B7,B8,,B3), using cell referencing. There is a double comma in the function's argument to indicate that the value in B3 is the future value, not the present value. This is consistent with Example 2.14 in Chapter 2. Entering the equation with cell references retains flexibility for the spreadsheet model, allowing the user to input a new value for the desired future amount, the annual interest rate, the number of years, or both, and the monthly payment needed to fund the stream of desired future value will be recalculated.

Discussion

To have a total of $707,790 that is required to supplement their Social Security benefits by $3,000 a month beginning at age 65 to be drawn for 25 years to age 90, assuming a constant rate of 2% annual growth compounded monthly, the couple should save $1,165 a month between the ages of 30 and 65.

What if our hypothetical couple didn't realize they needed to start saving for retirement until they were age 40 or even age 45? Running the same model again shows the monthly amount required to have the total of $707,790 by age 65 shoots from $1,165 a month starting to save at age 30 to $1,820 a month starting to save at age 40 and then to $2,401 a month starting to save at age 45, as shown in Table 5.3. These are easily calculated in Excel by simply changing t, the number of years in the spreadsheet model. These answers depend on stable interest rates over long periods of time, which may not be a reasonable assumption to hold. They also are made independent of any consideration of tax liability, whether funds were saved before or after taxes, all complications of the modeling that could be taken into account in a more complex modeling environment.

Table 5.3. *Future Value of Monthly Payments of $R Assuming r% = 0.02 Interest Over t = 25 and 20 Years*

	A	B	C	D
1	Future Value of a Stream of Payments			
2	Start saving age 40:		Start saving age 45:	
3	Desired Future Amount, S	$707,790	Desired Future Amount, S	$707,790
4	Annual interest rate, r	0.02	Annual interest rate, r	0.02
5	Number of years, t	25	Number of years, t	20
6				
7	Monthly interest rate, i	0.001666667	Monthly interest rate, i	0.001666667
8	Number of interest periods, n	300	Number of interest periods, n	240
9				
10	Monthly payment needed to fund	$1,820	Monthly payment needed to fund	$2,401

5.2   Managing Residential Mortgages

Part of the American dream includes owning a home. Since the housing market crash of 2008, that has been more difficult to achieve. Before the market bubble burst, many homeowners refinanced their homes, sometimes with variable-rate or even interest-only mortgages that encouraged home buyers into pricier homes, the prices of which fell further once the bubble burst. Housing short sales and foreclosures have dominated the headlines, even as mortgage rates have fallen in the past few years. New home buyers, doubly cautious, spend time investigating mortgage options before venturing into the housing market. Careful management of home mortgages is important for all home owners.

The staple of home mortgages is the fixed-rate home loan. Knowing how to build an amortization schedule is an easy, but important step in managing a home loan.

Some of the expenses of owning a home can affect federal and state income taxes. Specifically, the interest expenses for home mortgages are usually deductible, which lowers one's taxable income. Property taxes are often a deductible expense as well. So, let's make some assumptions and investigate the impacts of interest and property tax deductions by updating our amortization schedule.

Example 5.3: Building an Amortization Schedule

The Question

A couple plans to take out a fixed-rate home loan in the amount of $240,000 to fund the purchase of their new home. They locked in a fixed rate of 4.5% for a 30-year loan commitment. Their home is scheduled to close July 2, and the loan will fund July 1. Build an amortization schedule for their loan showing payments to interest and principal for the first 18 months of their loan.

Using Excel

Set up the headers in cells B1, B3, B4, and B5 through D5. Enter the value 240000 in cell D5. In cell A6, enter the value 1. In cell A7, enter the equation: =A6+1. Copy the equation from cell A7 down through cell A23. Highlight the block of cells B5 through D23, in the Home tab, select Cells Format, select Format Cells…, and select Currency. Click OK to format all cells in the table to show standard currency format, with a dollar sign, commas setting off the thousands place values and two places following the decimal.

The model is designed so the user enters the value highlighted: the annual interest rate, r, entered in cell B2. Enter the value 0.045 in cell B2. In cell D2, enter the equation: =B2/12. In cell B3, enter the equation: =ROUND(–PMT(B4,360,D5),2). The PMT function calculates the

Answer

Table 5.4. *Monthly Payments to Interest and Principal,*
r = 0.045

	A	B	C	D
1	Monthly Amortization Schedule			
2	r =	0.045	i =	0.00375
3	Payment =	$1,216.04		
4	Payment #	Interest	Principal	Balance
5				$240,000.00
6	1	$900.00	$316.04	$239,683.96
7	2	$898.81	$317.23	$239,366.73
8	3	$897.63	$318.41	$239,048.32
9	4	$896.43	$319.61	$238,728.71
10	5	$895.23	$320.81	$238,407.90
11	6	$894.03	$322.01	$238,085.89
12	7	$892.82	$323.22	$237,762.67
13	8	$891.61	$324.43	$237,438.24
14	9	$890.39	$325.65	$237,112.59
15	10	$889.17	$326.87	$236,785.72
16	11	$887.95	$328.09	$236,457.63
17	12	$886.72	$329.32	$236,128.31
18	13	$885.48	$330.56	$235,797.75
19	14	$884.24	$331.80	$235,465.95
20	15	$883.00	$333.04	$235,132.91
21	16	$881.75	$334.29	$234,798.62
22	17	$880.49	$335.55	$234,463.07
23	18	$879.24	$336.80	$234,126.27

required payment and the ROUND function rounds the required payment to the nearest penny.

In cell B6, enter the equation: =ROUND(D5*D$2,2). The use of the dollar sign locks the row reference so that when the equation is copied and pasted into other cells, the cell reference to the calculated monthly interest rate remains stable. In cell C6, enter the equation: =B$3–B6. The use of the dollar sign again locks the row reference so that when the equation is copied and pasted, the monthly payment remains stable. The ROUND function makes sure the interest charge is rounded to the nearest penny so that the model agrees with calculations typically done by the lender. In cell D6, enter the equation: =D5−C6. Highlight cells B6 through D6. With your mouse, grab the lower right-hand edge of the box where the solid square appears and drag that box down through cell D23. Excel will return the calculations captured in Table 5.4.

Discussion

The couple should plan for their monthly payment to the lender to be $1,216.04. For any given monthly payment, the values shown across that row reflect the portion of their total monthly payment that goes to interest, the portion that goes toward principal, and the outstanding balance of their mortgage.

Example 5.4: Updating an Amortization Schedule with Tax Savings

The Question

A couple plans to take out a fixed-rate home loan in the amount of $240,000 to fund the purchase of their new home. Assume the negotiated purchase price of the house was $300,000, and they plan to put $60,000 down on the home. They locked in a fixed rate of 4.5% for a 30-year loan commitment. Their home is scheduled to close July 2, and the loan will fund July 1. In addition to building an amortization schedule for their loan showing payments to interest and principal for the first 18 months of their loan, assume the couple will see 1.5% property taxes due on their purchase. To simplify the model, we will assume they will pay the property taxes due for the months of July through December in December and property taxes for the months of January through June in April of the same year. Assume the couple recognizes a combined federal and state tax rate of 30% on their incomes. Augment the amortization schedule with the cyclic payment of property taxes due each December. Finally, net the savings to their federal and state tax burdens from the mortgage and property tax payments made to produce an estimate of the real cost to them for buying the home the first year and a half.

Answer

Table 5.5. Augmented Amortization Schedule, Including Property Taxes and Income Tax Deductions

	A	B	C	D	E	F	G	H
1	Augumented Amortization Schedule							
2	r =	0.045	i =	0.00375				
3	Payment =		$1,216.04					
4	Payment #	Date	Interest	Principal	Balance	YEAR-TO-DATE SUMMARIES		
5					$240,000.00			
6	1	July, Yr 1	$900.00	$316.04	$239,683.96			
7	2	Aug, Yr 1	$898.81	$317.23	$239,366.73			
8	3	Sept, Yr 1	$897.63	$318.41	$239,048.32			
9	4	Oct, Yr 1	$896.43	$319.61	$238,728.71	YEAR 1		
10	5	Nov, Yr 1	$895.23	$320.81	$238,407.90	Property Taxes	Interest Paid	
11	6	Dec, Yr 1	$894.03	$322.01	$238,085.89	$2,250.00	$5,382.13	
12	7	Jan, Yr 2	$892.82	$323.22	$237,762.67	Year 1, Combined Interest & Property Tax Expenses		
13	8	Feb, Yr 2	$891.61	$324.43	$237,438.24			
14	9	Mar, Yr 2	$890.39	$325.65	$237,112.59		$7,632.13	
15	10	Apr, Yr 2	$889.17	$326.87	$236,785.72	Year 1, Savings from Income Tax Deductions		
16	11	May, Yr 2	$887.95	$328.09	$236,457.63		$2,289.64	
17	12	June, Yr 2	$886.72	$329.32	$236,128.31	NET EXPENSE, YEAR 1		$7,256.60
18	13	July, Yr 2	$885.48	$330.56	$235,797.75			
19	14	Aug, Yr 2	$884.24	$331.80	$235,465.95			
20	15	Sept, Yr 2	$883.00	$333.04	$235,132.91			
21	16	Oct, Yr 2	$881.75	$334.29	$234,798.62	YEAR 2		
22	17	Nov, Yr 2	$880.49	$335.55	$234,463.07	Property Taxes	Interest Paid	
23	18	Dec, Yr 2	$879.24	$336.80	$234,126.27	$4,500.00	$10,632.86	
24								
25						Year 2, Combined Interest & Property Tax Expenses		
26							$15,132.86	
27						Year 2, Savings from Income Tax Deductions		
28							$4,539.86	
29						NET EXPENSE, YEAR 2		$14,552.62

Using Excel

Augment the amortization schedule with a summary of property taxes and interest payments made during the calendar year. Year 1 includes just 6 months, so property taxes are calculated in cell F11 with the equation: =300000*0.015/2. The total interest payment made Year 1 in cell G11 is the sum of cells C6 through C11: =sum(C6:C11). To calculate the property taxes for Year 2, enter in cell F23 the equation: =300000*0.015. The total interest payment made Year 2 in cell G18 is the sum of cells C12 through C18: =sum(C12:C18). The combined interest and property income tax deductions for Year 1 in cell G14 is the sum of cell F11 plus cell G11: =F11+G11. Likewise, the combined interest and property income tax deductions for Year 2 in cell G26 is the sum of cell F23 plus cell G23: =F23+G23. The savings from income tax deductions for both years is 30% of the combined interest and property income tax deductions for each year. So, in cell G18, we include the equation: =G14*0.3; and in cell G28, we include the equation: =G26*0.3.

The net expense for holding the home Year 1 is the sum of the property tax for Year 1 in cell F11 plus 6 times the monthly mortgage payment of $1,216.04, calculated in cell C3, minus the savings from income tax deductions for Year 1 in cell G16. So in cell H17, enter the equation: =F11+6*C3–G16.

The net expense for holding the home Year 2 is the sum of the property tax for Year 2 in cell F23 plus 12 times the monthly mortgage payment of $1,216.04, calculated in cell C3, minus the savings from income tax deductions for Year 2 in cell G28. So in cell H29, enter the equation: =F23+12*C3–G28.

Discussion

Including property taxes that amount to $375 a month, funds required to cover the mortgage and the property taxes total nearly $1600 a month. But, when the deductions to the couple's income taxes are taken into account, the real cost of owning the home in the initial years of the mortgage is just over $1,200 a month. The expense of owning a home compared to rental expense incurred in renting or leasing alternate living space should be evaluated on the real dollars of the net expense of home ownership rather than the monthly costs of the mortgage and property taxes. More complex assumptions can be factored into the expanded model to bring greater reality into the projections made here.

Notes

Chapter 1

1. http://www.fueleconomy.gov
2. http://www.fueleconomy.gov
3. http://www.fueleconomy.gov
4. http://www.fueleconomy.gov
5. http://www.foodreference.com/html/f-chick-consp.html

Chapter 3

1. http://www.fueleconomy.gov
2. http://www.fueleconomy.gov
3. http://www.bts.gov/publications/national_transportation_statistics/, Table 1–16.
4. http://www.bts.gov/publications/national_transportation_statistics/, Table 1–16.
5. http://www.bts.gov/publications/national_transportation_statistics/, Table 1–16.
6. http://www.census.gov/prod/cen2010/briefs/c2010br-03.pdf
7. http://www.census.gov/prod/cen2010/briefs/c2010br-03.pdf
8. http://www.bts.gov/programs/airline_information/baggage_fees/
9. http://www.bts.gov/programs/airline_information/baggage_fees/
10. http://www.bts.gov/programs/airline_information/baggage_fees/
11. http://www.bts.gov/programs/airline_information/ontime_tables/2012-05/index.html
12. http://www.bts.gov/programs/airline_information/ontime_tables/2012-05/index.html
13. http://www.fueleconomy.gov

Index

Lightning Source UK Ltd.

9 781606 492802